Science Fair Success Using the Internet

Marc Alan Rosner

Enslow Publishers, Inc.

40 Industrial Road   PO Box 38
Box 398   Aldershot
Berkeley Heights, NJ 07922   Hants GU12 6BP
USA   UK

http://www.enslow.com

Dedication

I dedicate this book to my wife, Christine, who gives me constant inspiration and support. And I dedicate it to my daughter, Keira, for it is she who will truly harness and use technology in ways we are only beginning to imagine.

Library of Congress Cataloging-in-Publication Data

Rosner, Marc Alan.
    Science fair success using the Internet / Marc Alan Rosner.
       p. cm. — (Science fair success)
    Includes bibliographical references and index.
    Summary: Explains how to use Internet resources, including e-mailing experts and using search engines, to enhance science projects, with sample projects in biology, chemistry, physics, environment and earth science, and astronomy.
    ISBN 0-7660-1172-0
    1. Science projects—Data processing—Juvenile literature.
    2. Internet (Computer network)—Juvenile literature. [1. Science projects.
    2. Internet (Computer network)] I. Title. II. Series.
    Q182.3.R68   1999
    507.8—dc21   98-25945
                                                CIP
                                                  AC

Printed in the United States of America

10 9 8 7 6 5 4 3 2

To Our Readers:
All Internet addresses in this book were active and appropriate when we went to press. Any comments or suggestions can be sent by e-mail to Comments@enslow.com or to the address on the back cover.

Illustration Credits: Bill Arnett, p. 78; Carl Lischeske, III, p. 44; © Corel Corporation, pp. 7, 24, 40, 52, 66, 75, 90, 98; © DejaNews, Inc., p. 29; © The Electronic Desktop Project, 64; © The Franklin Institute Science Museum, p. 16; Immaculate Conception School, Somerville, New Jersey, USA, p. 106; Miami Museum of Science, p. 91; Raman Pfaff, pp. 99, 101; © The Regents of the University of California, p. 96; © TerraQuest, p. 71; © Wired Digital Inc., p. 25; © Yahoo! Inc., pp. 12, 17, 19.

Cover Illustration: The Stock Market

Contents

Introduction

Computers are useful tools if you have a science project to complete. In fact, scientists invented computers and the Internet. This book is about using the Internet to create an interesting, interactive science project. Of course, you can use your computer as a word processor or to download pictures. But you can also use the most modern and interesting Web sites for learning. You can communicate directly with scientists on the cutting edge of research.

Modern personal computers are many times more powerful than their ancestors. Computers arrived in homes and classrooms recently. In the late 1970s, you could place your telephone handset on a simple modem, and your computer would find a link to the outside world. Today, you can use a home computer with a modem to surf the World Wide Web. You can read text, view pictures, hear sounds, and watch video. You can interact with the many millions of other people connected to the Internet—all in a much faster and more sophisticated way than before.

The opening chapters of this book get you online with valuable research tips and strategies. Each of the following chapters covers a different area of science. The chapters start with an example of a science project that uses the Internet in some way. Then it lists some great Web sites pertaining to the chapter's topic.

In this book you will learn how to use the Internet to help you choose your science project and how to make it unique and personal. As you do your research, you will learn to be creative and make decisions. Internet-based research can make your project more in-depth and up-to-date. Your finished project will reveal important things about science and you.

Are you willing to spend time online? Do you want to build your technical skills? Are you committed to a serious work schedule over several weeks? If so, then read on. If you follow the suggestions in this book, you will be able to build a project that will make you, your teacher, and your family proud.

Chapter 1

Science Projects and the Internet

What exactly is the Internet? The Internet is many computers linked electronically. When you visit a Web site, send e-mail, or download a file, you are exchanging information between one computer and another. Government and university scientists invented the Internet in the late 1960s so that they could share research findings. In 1990, a group of Swiss researchers developed the World Wide Web, a method for viewing the information on one computer from another, using text and graphics. This change helped bring today's explosion in computer use. The number of people with Internet access at home more than doubled during 1996 and 1997, and continues to increase rapidly.

How Can I Use the Internet for My Project?

You can use the Internet to support your research. Instead of using the World Wide Web *as* your project, use it to *enhance* your project.

Examples of good Internet use include

- Gathering and analyzing data from a weather site over two weeks.
- Learning about animal species in a far-off rain forest so that you can design a food web.

Examples of poor Internet use include

- Printing many pages from a Web site and mounting them on poster board.
- Linking quotes from different sources with no explanation and no references.

Project Preparation

How do you complete a science project? How do you even start one? Intricate, thought-provoking assignments scare some people. Science projects are big jobs that have to be done over weeks or months. The trick is to break your project into little pieces. In choosing your project, start with what interests you.

Let's examine how Deirdre, an imaginary student, chooses and develops her project. Deirdre is an eighth-grade physical science student at Xavier Middle School in Sherman, Connecticut. She has been assigned a two-month project. It must relate to physics, chemistry, or earth science. She must follow a scientific method and make a project proposal. When it is approved, she must show her data and write a rough draft of her science project report. Then she must make a clear, final draft and enter it in her school's science fair in a presentable manner. She has these deadlines:

February 15: Project proposal due.
March 1: Show initial data.
March 15: Rough draft of report due.
April 1: Final draft of report due.
April 7: Enter project in science fair for judging.

Deirdre is not worried about this schedule. She will break her project into parts and meet each deadline. In choosing a project, she thought about her interests. She has always been fascinated by weather. When she was little, she witnessed a hurricane with her family in Florida. Ever since then, she has followed news reports about storms and unusual weather events. She likes to watch the weather change. She asks her parents many questions about the atmosphere and ocean. When they cannot give answers, she asks her teacher. She knows she wants to do something on this topic. She knows she can take measurements of temperature and humidity. She might even be able to predict the weather.

Deirdre has a computer she wants to use to support her research. Her first step will be to narrow her topic and make a proposal.

Before Deirdre begins her research, she will establish some safety rules that all students should follow. There are two sets of guidelines, one for Internet use and the other for the project itself.

Safe Computing

The Internet can put you in touch with strangers. Deirdre knows she must use good judgment. She will follow these rules:

- Do not give your last name, address, or telephone number to anyone.

- Do not name your school without permission from your teacher and parents.

- If anyone you contact acts in an inappropriate way, tell a parent or teacher immediately. The adult can report the occurrence to your Internet service provider and other authorities.

- Read and follow the rules of your online service.

- Never give your password to anyone.

- Never give out other private information such as your credit card number or social security number. Your Internet service provider already has this information, so if someone asks for this information again, chances are the person is pretending to be someone he or she is not.

- Use computer virus protection software to scan disks and downloads.

- Download files only when necessary.

Safe Experimenting

There is a chance Deirdre will handle equipment, chemicals, or glassware during her project. She wants to be safe, so she will wear protective clothing and work in a proper laboratory when necessary. Her teacher will advise her on safety issues. And Deirdre will follow these rules:

- Follow the instructions for conducting your project. Plan your work with a teacher, including safety procedures.

- Obey the rules of the classroom and laboratory.

- Wear safety goggles with side protection when handling chemicals, glass, or fast-moving objects.

- Handle chemicals in proper containers.

- Mix chemicals following proper instruction; never change quantities or substitute other chemicals.

- Use a fume hood and a well-ventilated laboratory when necessary.

- Be careful with electricity, especially when near water, and only experiment with electricity under the supervision of a knowledgeable adult.

- Do not eat or drink while experimenting.

- Review the operation of laboratory safety equipment such as eye wash and fire extinguishers.

- Report any laboratory accidents, spills, or injuries to a teacher.

- Take note when an experiment calls for adult supervision.

Logging on to the Information Highway

This book will not teach you step-by-step how to use the Internet. If you do not yet know how to browse the World Wide Web, you can get a book on that topic. When you are comfortable "surfing the Net," you will be prepared to continue.

Deirdre has a personal computer at home she will use for Internet access. Her school science lab and library also have computers with World Wide Web access. Her home computer has 32 megabytes of RAM, a 133-megahertz processor, a one-gigabyte hard drive, and a 28.8 kilobyte modem. (Lower numbers than these would make her work slower.) She subscribes to an online service that provides e-mail and a Web browser. A Web browser is a program that allows you to view graphics on the Internet.

Here are a few recommendations for using Netscape Navigator® and other browsers:

1. Use the most modern browser your computer can support. The newer the version, the better it will

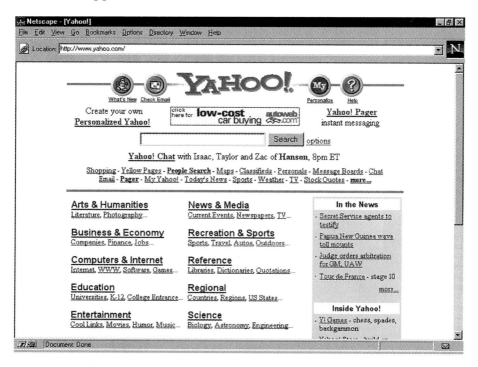

A web browser allows you to see graphics on the Internet. Yahoo! is one of many Web search engines that contain graphics.

support useful features such as Frames, Java, and Shockwave, some of which are called *plug-ins*. Many sites use these features, and more add them daily.

2. "Bookmark" useful sites so that you can visit them again without searching.

3. Highlight selected text in a browser window and paste it into a word-processing program such as Microsoft Word® or ClarisWorks®.

4. Try to work early in the morning, when Web traffic is low. The worst time is late afternoon and evening, when everyone wants to be online. At these times the Web may slow to a crawl.

5. Learn advanced features of your computer and software. On Netscape Navigator® and Internet Explorer®, you can link to the home page of the Web browser by clicking the button in the upper right-hand corner of the browser window. You can find support there. You can also learn to change the options in your browser with menu commands. Examples of advanced Web use you can learn on your own include

   • Increasing the memory allocation for your software.

   • Emptying the browser cache to speed things.

   • Configuring e-mail and newsgroups properly.

   • Downloading Web pages to your computer.

Choosing a Project

Deirdre is prepared to take the first step in her project: choosing a topic. She decides that the World Wide Web might be a useful place to start, so she visits some sites related to science projects to get ideas. The first one she goes to is:

Science Fairs Home Page

http://www.stemnet.nf.ca/~jbarron/scifair.html

This home page has listings of science project ideas at all levels and in all science areas.

Deirdre clicks on "Intermediate Projects" and finds project ideas for her grade level. She scrolls down the page and finds fifteen different project ideas under "meteorology," including

1. Snow—What happens when it melts; what it contains; structure of snowflakes; life in a snowbank.

2. Sky Color—Account for differences in color at different times.

3. Wind and Clouds—What are the common wind patterns in your area and why.

At this point Deirdre thinks it might be interesting if she recorded weather-related measurements at her home and tried to make predictions.

Many individuals and schools have made their project ideas available on the Web. Here are some other science project sites where you can find and share ideas:

Dr. Internet's Science Projects

http://www.ipl.org/youth/DrInternet/experiment.main.html

The Internet Public Library

http://www.ipl.org/youth/projectguide/

The Science Club

http://www.halcyon.com/sciclub/

Science Hobbyist
http://www.eskimo.com/~billb/

Virtual Science Fair
http://www.parkmaitland.org/sciencefair/index.html

There are also general Web sites for science education:

The Alive! Education Network: Science
http://alincom.com/educ/sci.htm

Canada's SchoolNet-Math and Sciences
http://www.schoolnet.ca/math_sci/

Cornell Theory Center Math and Science Gateway
http://www.tc.cornell.edu/Edu/MathSciGateway/

ExploraNet
http://www.exploratorium.edu/

Frank Potter's Science Gems
http://www-sci.lib.uci.edu/SEP/SEP.html

The Lab-ABC's Gateway to Online Science
http://www.abc.net.au/science/default.htm

SciEd: Science and Mathematics Education Resources
http://www-hpcc.astro.washington.edu/scied/science.html

Vicki Cobb's Science Page
http://www.vickicobb.com/

Each of these sites leads to resources in the different branches of science. If you want to see projects by other students, visit the following:

Kids Did This in Science!

http://sln.fi.edu/tfi/hotlists/kid-sci.html

> *This site links to student science projects in all the major branches of science.*

Searching for Useful Web Sites

Once Deirdre chooses her general topic, she is ready to do a search for useful Web sites. She starts by going to this search service:

Yahooligans! The Web Guide for Kids

http://www.yahooligans.com/

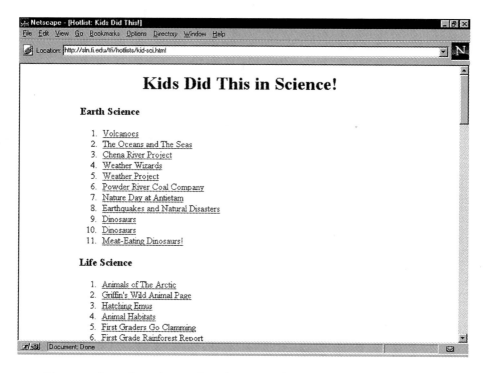

To see projects that other students have completed, visit the "Kids Did This in Science!" site. You may get an idea to start your own original project.

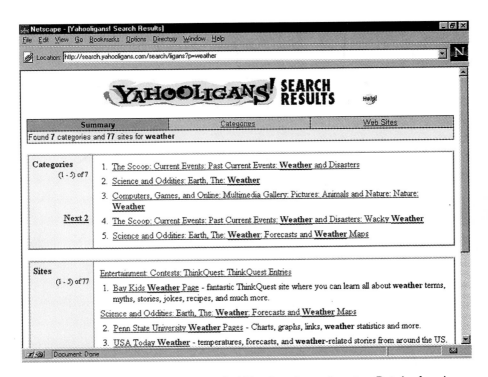

She enters the word "weather" and clicks the Search button. The computer provides her with seven categories and seventy-seven sites. She can visit those weather-related resources by clicking on anything that is hot text, that is, colored and underlined. The categories are groups of Web sites in the Yahoo! search service. The sites are direct links to Web resources on weather.

Deirdre uses the mouse to click on the suggestions from Yahooligans! She finds a wide range of weather sites. Some have weather forecasts, and others are about storms and natural disasters. Many suggest science experiments for students. A few are useful, but most are not. Deirdre understands that

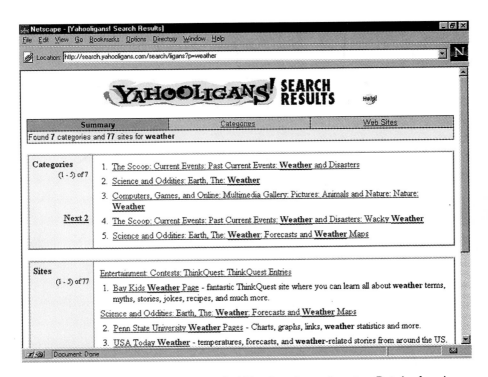

By entering the word *weather* into the Yahooligans! search engine, Deirdre found many sites that had relevant information she might be able to use for her science fair project.

people create Web sites for many reasons. Some sites are designed to share information, and others to sell products. Some are meant to express opinions. It is her job to decide which are relevant to her project. Her parents and teacher help her do this.

Deirdre notes which sites are useful. She bookmarks the good ones in the menu of her Web browser so that she can easily return to them later. She also makes a few notes with pen and paper so that she will remember what her favorite sites contain. This is a very important stage of research: sifting through all the information to get at what is useful.

It is important to understand that Yahooligans! is a search engine. *Search engines* are Web sites that list other sites. They allow you to enter a keyword to research a specific topic. The search engine does not contain the information you need, it merely points you in the right direction. Search engines are to the Internet what card catalogs are to libraries. Using both the Internet and the library will make your project stronger. Other search services you might visit are these:

Education World
http://www.education-world.com/

Disney Internet Guide
http://www.dig.com/

Study Web
http://www.studyweb.com/

The next search engines are more advanced because they contain many more sites. They also have advanced search features:

Infoseek
http://www.infoseek.com/

Webcrawler
http://www.webcrawler.com/

Yahoo!
http://www.yahoo.com/

Searching Tips

Here are some important tips for using search engines:

1. Read the instructions for each search engine.

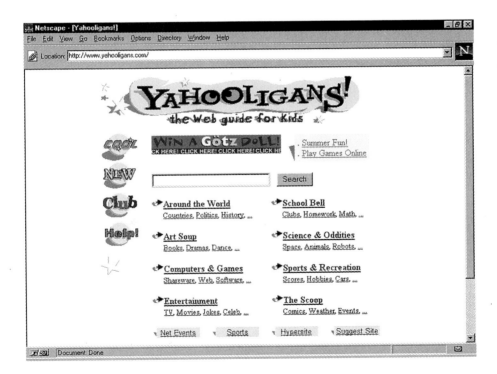

Yahooligans! is a search engine you can use to find Web sites that will help you research your science topic. To find out more about a search engine, click on the Help button.

2. Do not be afraid to surf around, checking out many sites before deciding which ones are useful. Make sure to visit several of the sites that come up.

3. Internet sites are not all the same quality. You can trust an Internet site hosted by the American Museum of Natural History more than one created by an America Online user with a personal account. In general, university and government Web sites are reliable.

4. If you find a useful site, bookmark it in your browser so that you do not lose it; it can be hard to find later if you forget how you got there. There are tens of millions of Web sites now!

5. If you do not find a useful site on your first attempt, or if you find too many sites, modify your query, using different word combinations. Deirdre's first search in Yahooligans! for "weather" produced seven categories and seventy-seven sites. When she searched for "weather events," she got two categories and seventeen sites.

6. Try different search engines to see which work best for you. They work differently. Some search engines list their sites because of human recommendations. Some search engines prowl the Web and find sites automatically that match the words you are searching for.

7. Learn the rules for each search engine. A search for *weather measurements* in Infoseek gave 1,630,240 pages. A search for "weather measurements" in quotes gave only 134 pages. The difference? The first search gave all Web sites it knew with the word *weather* or *measurements*. The second gave only those sites with that exact sequence of words, because quotation marks were used.

Conducting an Experiment: The Scientific Method

Deirdre wants to design and perform a good experiment. In doing this, she uses the scientific method. First she poses a problem or a question. Then she decides what kinds of experiments will help her find the answer. Using the results of her experiments, she draws a conclusion for her question.

Problem or question: What do you want to show? What questions do you want your project to answer? Deirdre decides on this question: "What factors determine the weather?" Often, the researcher follows the question with a hypothesis. A hypothesis is an explanation or a best guess at answering the question at hand. Then the experiment is conducted to try to prove the hypothesis.

Experimental method: What equipment and materials will you use to collect data, and how will you do it? Deirdre intends to take measurements the way a meteorologist would. She finds information on how to take measurements from weather maps and with other equipment at two sites:

Athena Curriculum: Weather
http://athena.wednet.edu/curric/weather/index.html

Weather Education-Precipitation Index
http://www.itl.net/Education/online/weather/precipind.html

Deirdre's method is to record temperature, air pressure, and dew point every day outside her home for four weeks. She makes her measurements in the same way at the same time each day.

Soon after starting her work, Deirdre arrives at this hypothesis: "A drop in air pressure leads to stormy weather." She forms this hypothesis from both her own data and from her Internet research.

Results: The results include your data, with relevant tables, graphs, and calculations. Since Deirdre has a Macintosh® computer, she decides to make her tables, graphs, and report by using ClarisWorks®. (A classmate of hers does the same thing with an IBM®-compatible computer by using Microsoft Office®.) She is able to cut and paste the tables and graphs into her report. She also uses her school's computer lab to print them out large and in color. She mounts the most important graphs on poster board for presentation in the science fair.

Conclusion: What does your research show? What would you do to continue your research if you had more time? Deirdre finds that when the air pressure drops suddenly over a twenty-four-hour period, the weather usually turns stormy. Sudden drops in air pressure come with wind and rain. This agrees with the information and theories she finds at general weather sites such as these:

Intellicast USA Weather
http://www.intellicast.com/weather/usa/

University of Michigan Weather Underground
http://groundhog.sprl.umich.edu/index.html

USA Today Weather
http://www.usatoday.com/weather/wfront.htm

The Weather Channel Home Page

http://www.weather.com/twc/homepage.twc

Deirdre follows the steps of the scientific method. She meets with her teacher a few times to show her the data she is collecting and to make sure she is on track. She spends time preparing her science fair entry. She uses large poster board to present her findings.

Of course, if your teacher provides a format for the scientific method that is somewhat different from this one, you should follow your teacher's format closely. Some projects, such as those that involve interviews or software programming, might have a different set of steps. The following sites will help with the scientific method. They offer tips on creating science projects, including the various stages (making observations, gathering information, and coming up with a title and hypothesis, etc.); sample projects; and presentation tips.

Experimental Science Projects: An Introductory-Level Guide

http://www.isd77.k12.mn.us/resources/cf/SciProjIntro.html

The Scientific Method

http://BugLady.clc.uc.edu/biology/bio104/sci_meth.htm

Other weather sites are listed on pages 64–65.

Chapter 2

Communicating with the Internet

The Internet is useful not only for the information it contains but also for the people with whom you will communicate. This chapter explores various ways to contact people online and gives you guidelines for online communication.

Forums and Message Boards

Forums and message boards are areas within online services where people send electronic messages about specific topics. There are forums on gardening, collecting coins, and investment; and there are forums on science education and science projects. Deirdre uses America Online (AOL) at home, so she did a keyword search on this service to find forums that relate to her weather project. She logged on to AOL and selected Find under the Go To menu. Then she clicked

"Find it on AOL." Then she entered "weather science projects," clicked Broad, and found that forty-nine areas came up in her search.

Some of the forums that came up were not useful to her. For instance, the Teacher's Lounge was a place for adult professionals, so she did not go there. *Weather News*, however, related directly to her project. It contained weather maps, forecasts, and data; weather-related discussions; and products for sale. Remember, much of the Internet is paid for by advertisers, so you will see ads come up in your research. You can ignore them unless they are useful in some way.

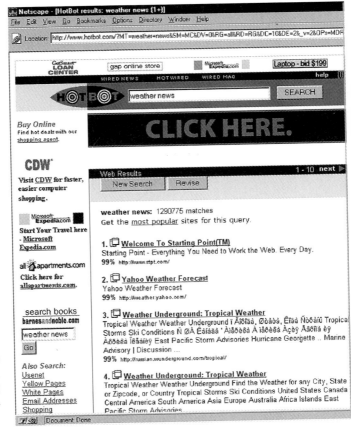

When you are searching the Web, you will notice paid advertisements for many products.

The Homework Help area allowed Deirdre to post questions and interact with teachers. Deirdre visited this area and decided to post a message asking for project help. First she read the instructions available there for doing this. Then she selected keyword "Ask-A-Teacher." Then she clicked on these items:

Junior High and High School ⇒

Post a Question ⇒

Earth and Life Sciences ⇒

Meteorology (Weather)

Before she posted her question, she noticed that there were 146 messages in the folder, and she spent some time reading them. To her surprise, some of her questions had already been asked and answered! After spending time browsing this area, she composed a question of her own and posted it:

> Subj: Weather Forecast Help
> Date: 2/20/99 7:02:23 PM
> From: DeirdreT85
> Hi,
>
> I'm doing a project for school on weather. I want to take measurements near my home and forecast the weather. I'm looking for suggestions about what measurements to take and how to analyze them. I know I want to measure temperature and rainfall, but what else? What patterns should I look for? I already bought a thermometer and rain gauge, and I'm studying weather on the World Wide Web. My project rough draft is due in about a month. Thank you.

Notice the strong points of Deirdre's request. She did not demand help, she asked politely. She was very specific in her request. She showed that she already started her research and is trying on her own. Imagine if she posted it this way:

Subj: Help me PLEEZ!
Date: 2/20/99 7:02:23 PM
From: DeirdreT85

Hey! I know lots of teachers and sceintists are out there. Can anyone tell me how to do my weather project? It's twenty five percent of my grade this term!

If you were a busy professional, which message would you answer?

Newsgroups

Newsgroups, also known as forums, bulletin boards, and round tables, bring together people who share the same interest. Newsgroup messages remain on the Web for all to see, even nonmembers. They are basically the largest electronic bulletin boards in the world. They are an effective research tool because you can see the record of electronic conversations that have taken place over time between interested participants. Check the Frequently Asked Questions file (FAQ) for your newsgroup. You may find that someone has already answered some of your questions. As with all your sources, cite newsgroups if you get information from them for your project.

You can find a directory of newsgroups through your online service or Web browser or on the World Wide Web at such sites as this one:

DejaNews

http://www.dejanews.com/

DejaNews is an excellent guide to newsgroups. It tells you what the newsgroups are and how to use them, and it allows you to search for newsgroups and messages by topic.

Here are some examples of newsgroups that relate to science.

NEWSGROUP	TOPIC OF DISCUSSION
bionet.plants	plant biology
sci.astro	astronomy
sci.energy	energy, science, and technology
sci.environment	environment and ecology
sci.geo.oceanography	oceanography, oceanology, and marine science

Deirdre was having difficulty figuring out how to measure humidity. She knew it was important to her project, so she went to DejaNews and did a keyword search for "measuring humidity."

Thirty-six messages came up. One in particular caught her eye. It was titled "Weather Station" and seemed to be related to her question. This posting came up when she clicked on the news posting:

Does anyone know where I can get some weather measuring equipment? I do not want anything fancy, just something basic. An all-in-one would be ideal, e.g. barometer, humidity meter, etc. Any ideas?

```
Netscape - [Deja News - Quick Search Results]                              _ | 5 | X
File  Edit  View  Go  Bookmarks  Options  Directory  Window  Help
Location: http://www.dejanews.com/dnquery.xp?QRY=measure+AND+humidity&ST=QS&CT=&DBS=1&defaultOp=%26&svcclass=dncurrent&maxhits=   N
```

Quick Search Results

Matches 1-20 of exactly 186 for search:

| measure AND humidity | Find |

- Help
- Power Search
- Interest Finder
- Browse Groups

Date	Scr	Subject	Newsgroup	Author
1. 98/07/20	036	Re: device to measure humidi	alt.home.repair	Rick Matthews
2. 98/07/19	036	Re: device to measure humidi	alt.home.repair	Laura Southard
3. 98/07/17	035	Re: device to measure humidi	alt.home.repair	CBress1
4. 98/07/17	035	Re: device to measure humidi	alt.home.repair	Des Bromilow
5. 98/07/17	035	Re: device to measure humidi	alt.home.repair	PQ
6. 98/07/16	035	device to measure humidity	alt.home.repair	Yong Huang
7. 98/07/21	034	Re: Humidity in House is Ver	alt.hvac	bristolracefan
8. 98/07/16	034	Re: device to measure humidi	alt.home.repair	Speedy Jim
9. 98/07/16	034	Re: device to measure humidi	alt.home.repair	shaun99
10. 98/07/16	034	Re: device to measure humidi	alt.home.repair	Karl Juul
11. 98/07/20	033	Re: Do clouds and humidity a	rec.radio.cb	odo
12. 98/07/16	033	Re: device to measure humidi	alt.home.repair	MJS
13. 98/07/16	033	Re: Humidity in House is Ver	alt.hvac	Greg
14. 98/07/17	031	Re: Having a good summer?	alt.religion.buddhism	Mr T
15. 98/07/16	030	Re: How did you store cigars	alt.smokers.cigars	ASGDC
16. 98/07/15	030	Re: Water immersion sensor	alt.engineering.elect	WarKang
17. 98/06/30	030	Re: heat index?(101)	alt.hahaha	Jim
18. 98/06/29	030	Humidity	alt.talk.weather	MOlson7924
19. 98/06/25	030	Re: 70% Humidity..Bah Humbug	alt.smokers.cigars	William E. Seit

```
Document: Done
```

Deirdre did a keyword search in DejaNews to help her find messages relating to measuring humidity.

She clicked on Next to see if anyone answered. A long message followed, some parts of which were useful:

On measuring humidity—Woodcraft (www.woodcraft.com) sells a hygrometer (about 2.5" diameter) for under six dollars. . . . The woodcraft hygrometer is good-looking and accurate, and just the right size (inexpensive too!). Note: The www.woodcraft.com site does not show the hygrometer, but there is a form where you can request their free print catalog. There are other sources for inexpensive accurate hygrometers—you do not have to pay fifteen or twenty dollars for one.

Deirdre used this information to shop for a humidity meter.

Posting Messages to Newsgroups

When looking for advice and help, you should first try forums on an online service. They tend to be well-organized areas with helpful people in charge. Newsgroups are appropriate if you need more advanced communication. An easy way to search for newsgroups and to join them is to go to America Online's area for this, using the keyword "Newsgroups." You can also click on "My DejaNews" at the DejaNews Web site to get instructions on joining and logging in to groups. As with forum and e-mail messages, you should learn the rules and culture of the newsgroup by watching it for a few days before posting messages. Read the FAQs (Frequently Asked Questions) first! Your tone should always be polite and businesslike. Members of mailing lists and newsgroups dislike when someone posts a message or ad not directly related to the theme of the group. This is called spamming, and it is impolite.

E-mail

E-mail stands for "electronic mail." It is a way of sending a written message to someone by computer. To do this you need an e-mail account, which you can get through your Internet service provider, school, or certain free Web sites such as <http://www.rocketmail.com>. E-mail is a good way to communicate in modern times. It is fast. It saves paper. You can store your messages and replies easily.

Sometimes people send e-mail without checking for proper grammar or spelling. Sometimes they are impolite. Do not make these mistakes. Do not send any message that you would be uncomfortable sending by regular mail or reading aloud in person. And always show your best side when sending mail to a respected professional or someone you do not know. If you act serious, you will be taken seriously.

Online services provide their own instructions for using e-mail. Here are some ideas on how you can use this powerful and modern method of communication to reach scientists who might help you with a project.

Contacting Scientists Online

One of the ways you can use the Internet in a science project is to reach scientists. Although scientists are very busy people who often work long hours and receive dozens of messages a day, they enjoy sharing their knowledge with interested students. It is important that you contact them only after you have worked on your project for a while. You want to get their attention and let them know that you have made an effort to learn on your own and that you have a genuine interest in their research. When should you ask a scientist a question? Ask yourself these questions first:

1. Was I unable to answer the question on my own, even after trying?

2. Is this person an expert who could provide specific information in this area?

If you answer yes to both questions, it might be worth it for you to e-mail this scientist.

If you find a Web site that puts you in touch with a scientist, carefully read the instructions before making contact. You may find that someone has already asked your question and the answer is posted online. If you still feel it would be worth it to ask your question, then go ahead—just be sure to compose it carefully.

Deirdre posed a question to a meteorologist she found on the Web:

Date: February 20, 1999 8:19:58 P.M.
Subj: Request for Science Project Assistance
To: PFeingold@weather.feinlab.com

Dear Dr. Feingold,
   I found your name in the Scientists-Helping-Students Web site. I hope you can take the time to answer a question I have. I'm trying to forecast my local weather. So far I have started measuring air temperature, pressure, and humidity every day. I know that a drop in air pressure can indicate bad weather. But how far does it have to drop, and how fast, for a storm to form? I could not find this information in my school library or on the Web. Since you are a meteorologist and do research, you might have more experience in this area. If I hear from you within two weeks, I can use your answer to support my project. Thank you for your time. I hope to find out more about what you do.

Sincerely,
Deirdre
e-mail: DeirdreT85@aol.com

To summarize, your message should be very specific and should contain the following important features:

- *Who* you are (but do not give your last name).

- *What* you want to know, and what you have already learned on your own.

- *Why* you are asking this person for help.

- *When* you need the information.

- A polite and professional tone.

It does *not* contain the following:

- A vague question.

- A demand for assistance.

- Errors in spelling or grammar.

- Insufficient time for the scientist to respond.

- Your telephone number or home address.

Here is another tip: People generally enjoy talking about their work. Ask them about their research, and they will usually tell you all about it. This will help with your research. And when you get a reply, the following are essential:

1. Thank them immediately and let them know they have helped you.

2. Cite them in your research as a source (see pages 37–40).

Here are some of the many Web sites where you can pose questions to scientists:

Ask an Engineer
http://www.city-net.com/~tombates/askengalt.html

Ask A Scientist
http://olbers.kent.edu/alcomed/Ask/ask.html

Dr. Odenwald's ASK THE ASTRONOMER
http://www2.ari.net/home/odenwald/qadir/qanda.html

The Last Word

http://www.newscientist.com/lastword/lastword.html

The MAD Scientist Network

http://medicine.wustl.edu/~ysp/MSN/MAD.SCI.html

OMSI Science Whatzit!

http://www.omsi.edu/online/whatzit/home.html

The Science Club

http://www.halcyon.com/sciclub/kidquest.html

Scientific American: Ask the Experts

http://www.sciam.com/askexpert/

SCORE Ask A Scientist

http://intergate.humboldt.k12.ca.us/score/noframes/ask.html

Society for Amateur Scientists

http://web2.thesphere.com/SAS/

The Why Files

http://whyfiles.news.wisc.edu/index.html

You are not limited to these sites. Many scientists have their own Web sites or work for laboratories or research institutions that have Web sites. You can find them by doing online searches or by contacting universities or government institutions. These Web sites almost always have an e-mail link that you can click on to send a message to one of the scientists or to the site administrator.

Mailing Lists

People join mailing lists with others who share an interest. This kind of list is called listserv. When a member sends a message to the listserv, it automatically goes out to all the members. There are mailing lists for all types of interests, including specific scientific fields of study. Every listserv has a list of FAQs that you should consult when you join. Think carefully before you send out messages to listservs. Some have thousands of members. An adult should approve any message you send. It is an excellent idea to monitor the listserv for a week or so before you send any messages, to learn the culture of the group.

Using listservs is an advanced Internet skill and should be done with help from an adult. If you feel you must join a listserv for your project, you can do a keyword search for mailing lists here:

E-Mail Discussion Groups

http://www.liszt.com/

Deirdre could have done a keyword search for "weather." It would have produced over a hundred mailing lists used by professionals.

A few e-mail discussion lists are designed for students in particular. This one, for instance, is described on the site just mentioned:

NYC-TO-AFRICA

This discussion list is dedicated to bringing together science teachers and students from New York City and Africa. You can mail a request for an "info" file to the server address, <listproc@lists.nyu.edu>. Put this single line in the mail message: info nyc-to-africa.

You can get a general "help" file from that server by mailing this single word: *help* to the same server address, <listproc@lists.nyu.edu>. This should also give you instructions on how to join or leave the group.

Electronic mailing lists are good for people who want to follow a topic in-depth over a long period of time.

E-Zines

E-zines are electronic magazines. New e-zines are launched each month, and many are free. One or two might be of interest to you. An example of an e-zine relating to science projects:

Cyberschool Magazine
http://www.cyberschoolmag.com

Cyberschool Magazine has featured articles in science, information on inventions and science research, and the Surfin' Librarian, which can help you find resources on the Internet.

A comprehensive and updated listing of e-zines can be found at this site:

John Labovitz's E-Zine List
http://www.meer.net/~johnl/e-zine-list/

FTP

File Transfer Protocol (FTP) is a method of transferring files between computers. If you find reference to a file on the Internet and it says you should obtain it using FTP, then follow the instructions of your Internet service provider for doing this. You can find those instructions in the "Internet" or "Member Services" area of your online service. (If you subscribe to America Online, use keyword "FTP.") Many universities have large numbers of files available for transfer. If you download files to your computer, you should have virus protection software on your computer. Some online services scan files for viruses before they make them available to users.

Cite Your References

As with any essay or term paper, it is essential for you to cite your research sources, whether you quote someone directly or rephrase their words or ideas. You probably already know that you must cite references to books and magazines, using footnotes and a bibliography—by presenting title, author, and publication information. But how do you cite the Internet?

Deirdre found a good quote on barometric pressure and hurricanes she wanted to use. Notice how she presented her information:

"In North America, barometric measurements at sea level seldom go below 29 inches of mercury (982 millibars), and in the tropics it is generally close to 30 inches (1,016 millibars) under normal conditions. Hurricanes drop the bottom out of those normal categories. The Labor Day hurricane that struck the Florida Keys in 1935 had a central pressure of only 26.35 inches (892 millibars). And the change is swift: Pressure may drop an inch (34 millibars) per mile." [Leon County Division of Emergency Management (LCDEM), 1997. "Hurricane Survival Guide," <http://www.co.leon.fl.us/lcem/anatom. htm> March 2, 1998].

She cites her source with the author (in this case an organization; sometimes it is a person), date of publication, title, Web address, and date she found the information.

To cite newsgroup and e-mail sources, give the author of the message, the date it was posted, subject of message, name of discussion list, and date you got it. For instance, look at how Deirdre presented her reply from Dr. Feingold:

"Dear Deirdre,
    In my experience, a drop of over 10 millibars per hour for more than three hours indicates a bad storm is on the way. Of course this depends on the time of year and the geographic location. Hurricanes are more likely to form in certain regions where they can be warmed by ocean water, such as coastal Florida. You may use actual data from our research station in your report. You will find it at <http://www.weather.feinlab. com/data/hurrcane.html>" [Dr. Terry Feingold, "Request for Science Project Assistance," e-mail dated February 20, 1999.]

If you want to learn more about how to cite software and Web sites, check these addresses:

Citing Web Resources
http://www.lafayette.edu/library/cite.html

Electronic Sources: APA Style of Citation
http://www.uvm.edu/~xli/reference/apa.html

The most important thing is that when you use someone else's work to do research, you give them credit. Give enough information for your reader to find your original source. Your teacher may give you a format to follow. Do not be afraid to ask for help. In general, your research will yield more reliable results when you use several Web sites, comparing and contrasting your findings, rather than simply taking quotes and images from one place.

Advanced Ideas

You can use programs such as Web Whacker® to download entire Web sites to a computer. This program allows you to display the site and all it contains later on, when you might not have Internet access.

If your computer has video-out capacity, you can capture animations and videos on a VCR.

At the completion of your project, you may want to create a Web site of your own to share with the world. Most online services have instructions on how to create Web pages. This can be part of your project.

More and more schools are getting modems and computer networks. You might be able to bring in your files on a floppy or zip disk. Or you might bring a laptop computer to school, with a security cable to keep it in place.

Chapter 3

Biology Projects Using the Internet

Huey was assigned a seventh-grade life sciences project. He chose to do a project on cells, the building blocks of tissue in plants and animals. Because cells are so small, scientists view them with microscopes. Optical microscopes use lenses. Other microscopes, such as electron microscopes, give different (and sometimes clearer) images of cells and cell parts. Viewing cells under a microscope in biology class, Huey found their appearance very interesting.

The main question Huey sought to answer was, how does the type of light source affect development of plant cells? Huey hypothesized that the plants would grow best under a light source like sunlight, which is a mixture of different wavelengths of radiation. He planned on growing cells under five different lighting conditions:

- fluorescent light

- incandescent light

- grow light

- sunlight

- darkness

Huey got the bulbs he needed from a lighting store. He grew two types of plants—peas and tomato plants—under each type of light. (He got his teacher's help in deciding which cells to use and how to prepare them.) He collected tissue samples from different stages of development, mounted them, stained them, and created labeled sketches of the cells and their parts. Through this method he showed that the plants with the most developed and healthy cells had grown under the light sources most similar to sunlight. The plants grown in darkness and under incandescent light had cells that were smaller and irregular in formation.

Huey wanted his project to show the structure and function of cell organelles, or cell parts. *Structure* refers to the composition, shape, and position of the organelle—what it *is*. *Function* is what the organelle does, how it keeps the cell alive and makes it work. Structure and function are very important to biologists, especially those who study cells, anatomy, and evolution. To make his project interesting and informative, Huey used different materials in his presentation:

- A working microscope with some cell samples.

- Sketches of cells and labeled diagrams, made by hand.

- Photographs of cells, taken with a camera that adapted to the microscope in his science class.

- Models of a plant cell made from common household materials such as cardboard, foam, and dried pasta.

- Images of cells from the World Wide Web.

Huey did not use the Web for every image of cells, only for those images that he had trouble producing on his own. For instance, his microscope was not powerful enough to show ribosomes, which are very small. He had to present them in a different way.

Huey did some Web searches for cells and cell-related terms. He used terms such as "cell," "ribosomes," "virtual cell," "organelles," and "nucleus," entering these keywords in the search engines. He found this site early in his search:

Cells . . . The Home Page

http://www.dcn.davis.ca.us/~carl/cellhome.htm

The opening page of this site provided the names of cell organelles. He viewed them by clicking on the names. For instance, when he clicked on the word *nucleus*, it brought him to a picture of that structure. It also described the function of the nucleus. He used hypertext ("hot") links to related structures such as "endoplasmic reticulum." He was able to wander around this site as if he were walking around the rooms of a house. He printed the color images he needed to fill out his report.

Huey found useful images at other sites, too. For example,

Virtual Cell

http://ampere.scale.uiuc.edu/~m-lexa/cell/cell.html

has electron microscope images. Electron microscopes are much more powerful than optical ones. They produce crisp black-and-white pictures. They are not really photographs, but images that suggest what the cell and organelle surfaces would look like if you could see them. This was useful to Huey in representing parts of the cells too small to observe with his microscope. He also used

Cells Alive
http://www.cellsalive.com/
which has illustrations and animations of cell processes.

Huey's project was successful. He showed the relationship between variable light and the development of plant cells. He identified some differences between animal and plant cells. For instance, plant cells have chlorophyll and animal cells do not. Plant cells also have more rigid cell walls. Huey used a few images from the Internet. Most of what he displayed, however, was of his own creation. He included his own sketches, photographs, and models.

Other Sites to Help You in Your Biology Research

Animal Anatomy

Cats
http://www.nationalgeographic.com/features/97/cats/
At this National Geographic Society site, you can learn about the anatomy of cats. You can study their muscular and skeletal structures, examine the way they sense things, and learn how they behave individually and in groups.

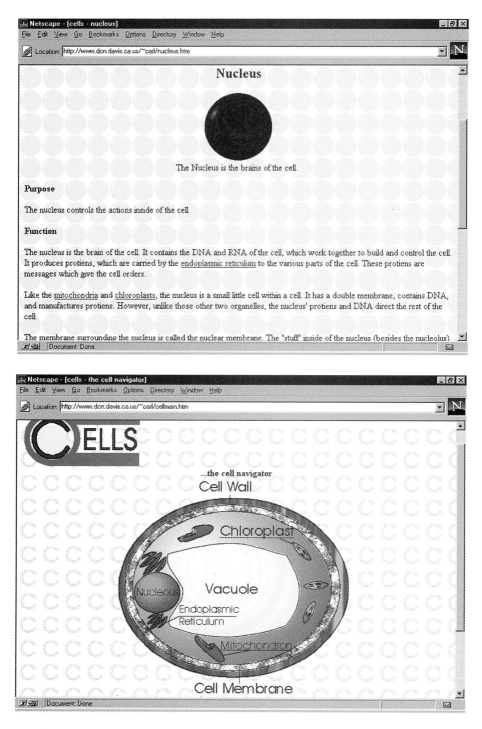

Images of cells and their parts can be found on the World Wide Web to enhance a cell biology project.

ChickScope

http://vizlab.beckman.uiuc.edu/chickscope/

At ChickScope, students operate an MRI (magnetic resonance imaging) machine over the Internet to view chicken embryos. The image of the egg can be viewed over the World Wide Web. You will be able to see "slices" of tissue and construct a picture of a developing embryo. This interactive Web site can support chick-hatching studies.

The Exploratorium's Cow's Eye Dissection

http://www.exploratorium.edu/learning_studio/cow_eye/index.html

At this site you can follow a virtual cow's eye dissection on the Web.

The Interactive Frog Dissection

http://curry.edschool.Virginia.EDU/go/frog

The Virtual Frog Dissection Kit

http://george.lbl.gov/ITG.hm.pg.docs/dissect/info.html

These two sites allow you to "dissect" a virtual frog. You can learn about anatomy without actually harming a frog. You can slice open your digital frog and examine the tissues and organs of these amphibians.

Whole Frog Project

http://www-itg.lbl.gov/Frog/

This is the main Web site for the frog project, where you can examine images and movies of frogs and frog parts in virtual reality.

Animal Studies

Conchologist's Information Network

http://erato.acnatsci.org:80/conchnet/

> If you collect shells, you will enjoy this site. You can identify different types of marine creatures here. There is a detailed reference for gastropod information.

The Gator Hole

http://magicnet.net/~mgodwin/

> Alligators are feared and often misunderstood by humans. At The Gator Hole, you can set the record straight by separating the myths from the facts. Learn about alligator biology: their anatomy, feeding habits, and reproductive methods.

Hide & Seek Overview

http://www.muohio.edu/dragonfly/hide/index.htmlx

> These Dragonfly activities deal with camouflage strategies. Camouflage is an important animal survival strategy in the wild.

How Wolves Communicate

http://www.muohio.edu/dragonfly/com/

> This site, also from Dragonfly, analyzes the way wolves communicate in a pack. It describes their strategies for establishing roles and marking territory. Then it leads to a way to analyze your own pets' behavior.

Iowa State University Entomology Image Gallery

http://www.ent.iastate.edu/imagegallery/

This site has many images from a large collection of insects. If you are interested in smaller creatures, learn about beetles, lice, butterflies, moths, cicadas, leafhoppers, and other insects here. You can collect insects from your home area, preserve them in clear containers of rubbing alcohol, and use printouts from the image gallery to enhance your study. Science stores sell insect containers or kits with built-in magnifying lenses.

Monarch Watch

http://www.MonarchWatch.org/

This award-winning Web site teaches about the world of Monarch butterflies. You can do a study of their ecological niche and migration patterns. There is also information on milkweed plants; use the site's handy photo guide to identify milkweed in your area, or find out how to grow your own plants.

National Zoological Park Home Page

http://www.si.edu/natzoo/

Naturalia

http://www.edv.it/naturalia.html

You can use these sites to support animal study. They can help you create a profile of the animal: What does it eat? Where does it sleep? What predators should it avoid? What strategies does it use to protect itself? Is it endangered? How has it adapted to the presence of humans?

The Nutty Hatch

http://www.geocities.com/Yosemite/7727/

This site is all about birds and bird-watching. You can learn about the different species of birds in the "gallery." You can build a nest box from the instructions there and learn how to feed and identify birds in your own yard.

USGS Biological Resources

http://www.nbs.gov/

This is another site for animal enthusiasts. It will direct you to libraries, organizations, projects and programs, fact sheets, current research, photos, stories, and educational areas.

USGS Patuxent Wildlife Research Center

http://www.pwrc.nbs.gov/

The Patuxent Wildlife Research Center was founded in 1936 as America's first wildlife experiment station and research refuge. You can get involved in the specific monitoring of amphibians, birds, butterflies, and other species at this site.

WhaleNet at Wheelock College, Boston

http://whale.wheelock.edu/

Whale Songs

http://whales.ot.com/

These two whale sites were designed for students. The second site contains journal entries from a science teacher studying whales in the ocean. You can read daily journal entries and listen to actual sounds produced by the world's largest mammals.

Wonders of the Seas

http://www.oceanicresearch.org/lesson.html

This is a growing collection of lessons on sponges, mollusks, and other creatures of the sea. It contains facts and diagrams describing the anatomy of these organisms, where they live, how they feed, and what strategies they use to survive.

Ecology

Bugs in the News!

http://falcon.cc.ukans.edu/~jbrown/bugs.html

Bugs in the News! is not about insects but about viruses, bacteria, and other microorganisms such as E. coli. E. coli *is a helpful bacterium when it is in our large intestines: it helps us digest food. However, when it contaminates meat or local water supplies,* E. coli *causes health problems.*

Digital Learning Center for Microbial Ecology

http://commtechlab.msu.edu/sites/dlc-me/

Learn about the "microbe of the week" featured at this site. Visit the microbe zoo, with microbe-rich environments at "dirtland," "the snack bar," and other buggy spots.

Dragonfly Trees Activities

http://miavx1.muohio.edu/~dragonfly/trees.HTMLX

Here you can learn about trees by engaging in different activities. Topics include tree shapes, seeds, and how trees protect their space.

Fun Facts About Fungi

http://www.herb.lsa.umich.edu/kidpage/

This catalog explains fungi: what they are, where they are found, of what use they are to people. You can do a collection and presentation on fungi, enhanced by this site. Look for fungi in shady, moist areas, especially following rainy periods. Put them on display in sealed containers and describe how they grow. Did you find them in a grassy area? On the bark of a rotting tree? Under a rock overhang?

Illusions and Perception

Blind Spots
http://serendip.brynmawr.edu/~pgrobste/blindspot1.html

Illusionworks
http://www.illusionworks.com/

Optical Illusions
http://www.scri.fsu.edu/~dennisl/CMS/activity/optical.html
If you are interested in blind spots and optical illusions, visit these Web sites and try the experiments there. You will learn about how the eye perceives shapes and colors and how it delivers information to the brain through the optic nerve. You can compare different people's reactions to optical illusions. Do certain people experience them more than others? Is visual perception related to age or gender?

Human Anatomy

The Heart: An Online Exploration
http://sln.fi.edu/biosci/biosci.html

At this Web site you can study the structure of the human heart. You can wander around vessels as if you were a blood cell. Learn how people discovered things about the heart through past research.

Human Anatomy Online
http://www.innerbody.com/

This site is a fun, interactive collection of illustrations of the human body with animations and thousands of descriptive links. Human Anatomy Online uses Java applets to show images and select anatomy parts. You can focus on one particular anatomical system, such as the skeletal system, the nervous system, or the reproductive system.

Neuroscience For Kids
http://weber.u.washington.edu/~chudler/neurok.html

This home page has been created for elementary- and secondary-school students and teachers who would like to learn more about the nervous system. It contains activities and experiments relating to the brain and spinal cord.

The Visible Human
http://www.madsci.org/~lynn/VH/

In the Visible Human Project, scientists took slices of frozen human tissue and made more than eighteen thousand images. At this site, you can view the images in great detail from actual photos and videos. This project will help you understand a three-dimensional picture of the human body.

Chapter 4

Earth Science Projects Using the Internet

Susan wanted to do her earth science project on the earth's interior. She lives in San Francisco, where there are many minor earthquakes and sometimes major ones. The earthquakes made her curious about the structure of the earth. What is inside it? What would you find if you kept digging? How does the earth change over time? Why do certain areas of the world have more earthquakes than others?

Susan found earthquake-related sites in the search service Infoseek:

Infoseek Earthquake Search
http://www.infoseek.com/earthquakes

She remembered the 1989 earthquake in Loma Prieta, California. It was the first major event on the San Andreas fault since the infamous earthquake of 1906. She learned this and more at the following site:

UC Berkeley Seismological Laboratory FAQs
http://www.seismo.berkeley.edu/seismo/faq/1989_0.html

Susan wanted to compare this earthquake with an earthquake in New York to see why quakes are so violent in the region in which she lives. She looked up the New York earthquake nearest in time to the Loma Prieta one at this site:

Area Earthquakes
http://www.syzygyjob.com/areaquak.shtml

In 1985 an earthquake in Newburgh, New York, registered 4.0 on the Richter scale. The Richter scale is a measure of earthquake intensity. The higher the number, the more violent the earthquake. The Newburgh earthquake was a significant event, but nothing like the 7.1-magnitude quake of Loma Prieta.

Susan continued her research to learn more about the forces that shape the earth. She visited these sites:

Earth's Interior & Plate Tectonics
http://bang.lanl.gov/solarsys/portug/earthint.htm

EARTHFORCE
http://sln.fi.edu/earth/earth.html

The first site showed cross sections of the earth, and it explained what is inside it. Susan learned that the earth has a nickel-iron core surrounded by the mantle and then the crust. Different regions within these layers have their own names and properties. She also learned about the boundaries of ocean plates. Portions of the earth's crust, called plates, slide around. This causes earthquakes at the boundaries and has led to the position and shape of our modern continents.

Susan began to form a hypothesis, that earthquakes occur in regions where one crustal plate meets another. The rubbing of plates against each other, pushing and pulling, could be the source of earthquake energy. Perhaps California has more numerous and energetic earthquakes than New York, she theorized, because it is closer to the edge of a crustal plate.

The second site was an online exhibit provided by The Franklin Institute Science Museum. It explained earth forces in the crustal plates and at the ocean bottom. It also described tsunamis, which are giant tidal waves that result from earthquakes. Tsunamis have killed many people throughout history.

Marine Geology

http://walrus.wr.usgs.gov/docs/margeol.html

At this site, Susan learned more about the ocean in particular. It explained how scientists study the ocean floor. Marine geologists teach us how the earth got its present geography. They also help us with more practical problems: where to look for minerals, oil, and other resources.

MTU Volcanoes Page

http://www.geo.mtu.edu/volcanoes/

The Michigan Technological University shares its information about volcanoes. It describes, with maps, where they are found and explains how scientists study them. It lists terminology and definitions.

From these various sites, Susan learned that San Francisco is located at the boundary of the Pacific plate. Indeed, the entire West Coast of the United States is an area active with volcanoes and earthquakes. The Rocky Mountain range extends from Canada to Mexico. New York, on the other hand, is not at the edge of a crustal plate. Its smaller mountains are much older and have eroded over time.

Susan wanted to show how the crustal plates of the earth move. She wanted to communicate the type of processes that have determined the earth's geologic history. If she could model the movement of the earth's plates, the eruption of a volcano, and the effects of an earthquake, she could show how the plates generate heat. Such activity causes earthquakes and volcanoes, which cause the earth to change over time.

Geologists believe that the continents once formed a large landmass called Pangaea, and that these landmasses slid apart over the earth's fluid interior to form today's continents. Susan made an exhibit for her science project to demonstrate this. She went to the Earth's Interior & Plate Tectonics site (page 53) and dragged the Crustal Plate Boundaries file to the desktop of her computer. She opened this new "plates.gif" file in a paint program and enlarged it by 400 percent. She printed out the image and pasted it on thick cardboard. Then she cut the continents out, mounted magnets on the back of them, and put them in an iron tray. Visitors to her exhibit could move the continents together, like pieces of a puzzle, to form Pangaea. Then they could pull them apart to their modern positions.

Susan also visited:

Volcano World
http://volcano.und.edu

to look for ways to build models of a volcano. First she clicked on "Volcano Starting Points." Then she saw "Search All Documents on VW." She entered the word "model" and clicked the Search button. The first dozen or so links that came back did not seem very useful. But number eighteen was. It was called "Building Volcano Models Paper and Cardboard . . ." She clicked there, and the page that came up was just what she was looking for. It showed eleven different ways to make models of volcanoes! Not all were practical for Susan. She picked number three, simple clay models, because she had several colors of clay at home.

Since Susan got images and ideas from these Web sites, she made sure to cite them in her references. Her citations looked like this:

Rosanna L. Hamilton, (1995–1997), Earth's Interior & Plate Tectonics <http://bang.lanl.gov/solarsys/portug/earthint.htm> (May 9, 1999).

(Author & Date Unspecified), Simple Clay Models, Volcano World <http://volcano.und.nodak.edu/vwdocs/volc_models/ clay. html> (May 11, 1999).

Where possible, she cited the author, original publication date, name of Web page, address, and date she visited. Sometimes it is hard to figure out all this information from Web sites. Not all information is always present. Susan did the best she could to document her sources.

Finally, Susan was confused about the difference between S and P waves in earthquakes. She found someone to answer her questions at this site:

Ask-A-Geologist

http://geology.usgs.gov/ask-a-geo.html

She e-mailed her question to an actual USGS earth scientist and received an answer within a few days.

With all the information she had gathered, Susan was able to show clearly that regions of earth activity occur at plate boundaries. Earthquakes, volcanoes, and plate edges are clustered in the same places around the world. Areas between plate boundaries have much less seismic and volcanic activity.

Other Sites to Help You in Your Earth Science Research

Earth Science and Geology

Careers in Geoscience
http://www.science.uwaterloo.ca/earth/geoscience/careers.html
At this site you can find out how people become geoscientists. Who are today's geoscientists? What do they do? What can you do with this training? The site answers these and other career questions.

Earth Sciences
http://www-hpcc.astro.washington.edu/scied/earthindex.html
This is a general index site that leads to Web sites on geology and earth systems, meteorology, oceanography and marine biology, and environmental sciences.

Map-It: Form-Based Simple Map Generator
http://crusty.er.usgs.gov/mapit/
This site lets you enter longitude and latitude values. Then you click on a button to get a projection map of the region you indicated.

U.S. Geological Survey Learning Web

http://www.usgs.gov/education/

This is the USGS education site, with plenty of resources for students. It teaches about glacier movement, sea-floor spreading, arctic delta formation, landslides, and fossil formation. It also explains mapmaking processes.

USGS Radon Information

http://sedwww.cr.usgs.gov:8080/radon/radonhome.html

This site tells you about radon, a noble gas that can be a health hazard in homes. You can learn about where radon is located and how it is measured. Is your home or community at risk for radon?

Earth History

American Museum of Natural History: Fossil Halls

http://www.amnh.org/Exhibition/Fossil_Halls/index.html

London Natural History Museum Department of Palaeontology Home Page

http://www.nhm.ac.uk/palaeontology/

Royall Tyrrell Museum

http://tyrrell.magtech.ab.ca/home.html

UC Museum of Paleontology, Geology and Geologic Time

http://www.ucmp.berkeley.edu/exhibit/geology.html

Welcome to the Paleontological Research Institution

http://www.englib.cornell.edu/pri/

> *These sites describe the history of the earth. They share information about plants, animals, and geologic events. They also display fossils, teaching about the evolution of species and earth's changing geology.*

Earth Imaging Systems

GOES Project

http://rds.gsfc.nasa.gov/goesb/chesters/web/goesproject. html

JPL Mission to Planet Earth

http://www.jpl.nasa.gov/earth/

NGDC Science for Society

http://www.ngdc.noaa.gov/ngdc/ngdcsociety.html

TOPEX/Poseidon

http://topex-www.jpl.nasa.gov/

> *Learn how scientists use satellites and radar to understand climate and weather. They use imaging instruments to collect data about the ocean floor, wildfires, the climate, earthquakes, sea level changes, and other topics. You can download images, data, and video clips from the Web to do your own research.*

Minerals and Other Natural Resources

Mineral Exhibits

http://geology.wisc.edu/~museum/minerals.html

This site lists common minerals, many of which you can find near your home. You can collect, identify, and classify minerals. (Many science stores such as The Discovery Channel store and The Nature Company sell hard-to-find specimens.) This site has an especially good exhibit on fluorescent rocks and minerals that glow under ultraviolet light. Your teacher may be able to provide you with fluorescent mineral samples and a mineral light.

Smithsonian Gem & Mineral Collection

http://galaxy.einet.net/images/gems/gems-icons.html

This site has images of minerals you might not find near your home.

Virtual Cave

http://www.goodearth.com/virtcave.html

Are you a spelunker? You can view good-quality photographs of different mineral samples gathered from caves all around the world at this site.

Paleontology and Fossil Records

Middle School Earth Science Explorer

http://www.cotf.edu/ete/modules/msese/explorer.html

You can learn about earth systems and dinosaurs. This site explores some of the theories of why the dinosaurs became extinct so suddenly. One theory is that a giant asteroid crashed into the earth. Huge amounts of water and ash went into the atmosphere following the explosive impact, blocking the sun and cooling the climate for years.

Skeletons

http://www.muohio.edu/dragonfly/skeletons/index.htmlx

This site has a simple explanation of how fossils form. It leads to a "virtual dinosaur dig," where you pick the tools and strategies for finding ancient reptilian remains.

What Is a Fossil and How Is It Preserved?

http://www.pa.msu.edu/~sciencet/ask_st/082097.html

This scientist's answer to the question tells about the different types of fossil formation. Fossils can be formed by preservation, mineral preservation, casts, or imprints (such as footprints).

Volcano-Related Web Sites

If volcanoes interest you, there are many Web sites you can visit.

Cascades Volcano Observatory

http://vulcan.wr.usgs.gov/home.html

This Web site describes volcanoes in the American Pacific Northwest. It has many maps and other graphic images you can use. You could build a working model of an instrument used by seismologists or volcanologists. This would be useful for a project on human safety. If you live near an active volcano, you could create a volcanic eruption detection plan and emergency evacuation plan.

Dante II Frame Walking Robot

http://img.arc.nasa.gov/Dante/

When scientists want to study an active volcano that is too dangerous for humans to enter, they can send in a robot: Dante II. Dante bravely ventured into Alaska's Mount Spurr to get readings of gas fumes. This technology has uses in areas other than volcano research. Robots like Dante can be used to clean up hazardous waste sites or to diffuse bombs.

NASA EOS IDS Volcanology Team

http://www.geo.mtu.edu/eos/

The National Aeronautics and Space Administration (NASA) helps volcanologists by using a technology called remote sensing. This site explains how the process is done. It shares information discovered by government-funded research.

Volcano Watch

http://hvo.wr.usgs.gov/volcanowatch/

This weekly newsletter comes from volcano country: Hawaii. It focuses on the volcanoes of Hawaii and gives updates on their activity. Produced by scientists at the United States Geological Survey's Hawaiian Volcano Observatory, it contains a library of articles with a search engine.

Volcano World

http://volcano.und.nodak.edu/

This Web site features a volcano of the week. You can find out which volcanoes are erupting where. You can view images and videos of volcanoes in action.

Weather

In Chapter 1, we looked at how Deirdre used several weather sites to prepare her project. Here are some other weather-related sites:

El Niño: Hot Air Over Hot Water

http://sln.fi.edu/weather/nino/nino.html

> *El Niño is a huge patch of warm Pacific water that affects the climate in North America. At this site, The Franklin Institute Science Museum explains what El Niño is. There are activities you can do to demonstrate the thermal processes at work.*

Hurricane

http://www.miamisci.org/hurricane/

> *This site, from the Miami Museum of Science, describes the inside of a hurricane. You can learn about the instruments meteorologists use to make physical measurements during these violent storms. Family members share their frightening stories from actual storms. Southern Florida is a region that gets hit with many hurricanes.*

Ice & Snow

http://miavx1.muohio.edu/~dragonfly/snow/

> *You can visit Antarctica, learn about snow and ice, and experiment with virtual snowflakes.*

National Weather Service

http://www.nws.noaa.gov/

> *The National Weather Service is a government organization in charge of warning people about coming*

weather events. It provides information to weather bureaus at newspapers and television and radio stations. This helps people prepare for blizzards, heat waves, and other storms and important meteorological events.

NCEP Climate Prediction Center

http://www.nnic.noaa.gov/cpc/

This is another online resource for El Niño. At NCEP Climate Prediction Center, you can get pictures that contain temperature, wind, and precipitation data resulting from El Niño. This page also has a "U.S. Threats" section that predicts bad weather.

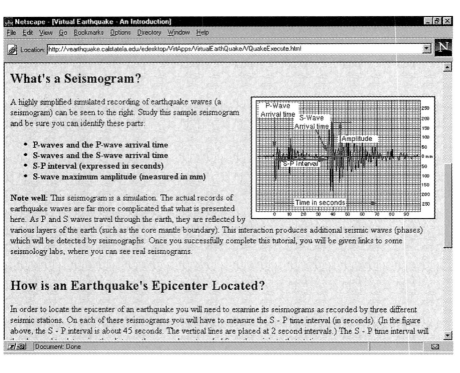

Advanced earthquake information can be found at the Virtual Earthquake site.

Advanced Earthquake Sites

Surfing the Internet for Earthquake Data

http://www.geophys.washington.edu/seismosurfing.html

> *Use this site as a home base to gather and analyze earthquake data. Who knows, maybe you can predict the next major earthquake!*

Virtual Earthquake

http://vearthquake.calstatela.edu/

> *The Virtual Earthquake site is good for advanced high school students. It allows you to find the epicenter of a mythical earthquake.*

When an earthquake occurs, the ground moves, and this causes vibrations and sound waves that travel through the earth. These waves move outward from the center of the earthquake and travel away. The more time that passes, the farther they travel, like waves in the ocean. Seismologists use science and math to figure out the starting point of the vibrations.

Chapter 5

Environmental Science Projects Using the Internet

Hydrology is the study of water in its various forms. Water can be a liquid, solid, or gas (vapor). It can be fresh or contain salt. It can be clean or polluted. Every known life form uses water to survive. Humans rely on the fresh, clean version of this precious commodity. Less than one percent of water on Earth is available as fresh and unfrozen. As the human population grows and reaches the remotest areas of our globe, we have to protect this natural resource ever more carefully.

Brendan's school was preparing for its annual science fair. Brendan wanted to enter a project in the seventh-grade environmental sciences category. He lived on a farm with a pond and a stream and decided it would be perfect if he could do a project on the hydrology of his home environment. He knew that these water bodies were not as healthy as they once had been: His father told him the water used to be clearer and the fishing used to be better. Brendan was interested in finding out what was harming the ecology and what could be done about it. He had heard something about eutrophication and low oxygen levels.

Brendan got Web site recommendations from various sources. He tried a Web search on his own for "water," "hydrology," "stream," and other terms. As it turned out, there is an entire section of Yahoo! devoted to hydrology:

Yahoo!: Hydrology

http://www.yahoo.com/science/earth_sciences/hydrology/index.html

This area of Yahoo! led Brendan to many water-related sites.

Brendan then asked his teacher if she knew of any hydrology sites. She showed him a magazine article that had environmental education Web site reviews. He visited these sites first:

Ocean Planet

http://seawifs.gsfc.nasa.gov/ocean_planet.html

The Ocean Planet site gave Brendan lots of information about the types of measurements scientists take on water. Brendan started to understand the water cycle on the earth, which was valuable background information for his project.

USGS Water Resources of the United States

http://water.usgs.gov/

This site has real-time data, meaning that it has actual statistics for present water conditions. It also has historical data and links to other water-related sites. However, Brendan was most excited about one thing the site offered: free posters. He wrote to the United States Geological Survey (USGS), and they sent him posters on wetlands, water use, groundwater, wastewater, and water quality.

Brendan did a formal study of the bodies of water on his farm. He got a test kit for measuring oxygen. With his teacher's help, he was able to measure the oxygen in the water and compare it with the oxygen levels in more balanced bodies of water. He found that oxygen, a life-sustaining gas, was indeed low in the water on his farm. He also monitored temperature, depth, and velocity. He studied the wildlife, making observations on the fish, mammals, amphibians, and plant life in the stream and pond. This part of the project really supported his hypothesis—that fertilizer was getting into the water, causing algae to grow and use up all the oxygen and nutrients needed by plants and fish.

For the final part of his project, Brendan made environmental recommendations for the protection and preservation of the hydrologic environment on his farm. The following site was useful for this part of his research:

National Marine Fisheries Service
http://kingfish.ssp.nmfs.gov/
This governmental Web site contains links to actual written passages from the Endangered Species Act.

Brendan concluded that his farm, and neighboring farms, could make some operating changes that would improve the health of the stream. He suggested that a change in the type of fertilizer used and the way it was applied might help restore the ecological balance of the water. By making some simple modifications in their techniques, the farmers might be able to reduce their effects on aquatic life. Brendan shared his recommendations with his father and with neighboring farmers.

Other Sites to Help You in Your Environmental Science Research

Ecological Environments

The World Wide Web brings places to you that you might not have otherwise seen. At the following sites, you can study remote ecological environments from home and school.

Desert Life

http://www.desertusa.com/life.html

Desert Plant Survival

http://www.desertusa.com/du_plantsurv.html

Rainforest Action Network

http://www.ran.org/ran/

> *At this site you can learn about the rain forest and find out how to help preserve these valuable areas of our environment. The kids' corner has a calendar contest and a question and answer section.*

Tropical Rain Forest In Suriname

http://www.euronet.nl/users/mblecker/suriname/suri-eng.html

Virtual Antarctica

http://www.terraquest.com/antarctica/index.html

These well-crafted sites examine particular places on our planet. You can study different environments at each. The sites describe climate, plants and animals (flora and fauna), food sources, human culture, and conservation.

Webs of Life

http://www.muohio.edu/dragonfly/webs/

At this site you can explore the islands of Baja and learn about the ecological niche of spiders. "Backyard Islands" will lead you to learning about your own home environment.

Global Forums and Projects

Alaska Science Forum

http://www.gi.alaska.edu/ScienceForum/index.html

Do you have a yearning to visit Alaska? This Web page is provided as a public service by the Geophysical Institute, University of Alaska Fairbanks. It leads you to Alaska-related resources in every science you can think of, from agriculture to zoology. Visit the science forum or go on science writer Ned Rozell's Pipeline Trek.

The Jason Project

http://www.jasonproject.org/

Every year, The Jason Foundation for Education sponsors an annual scientific expedition. Using advanced telecommunications, students can take part in live, interactive programs following the expedition. Visit this

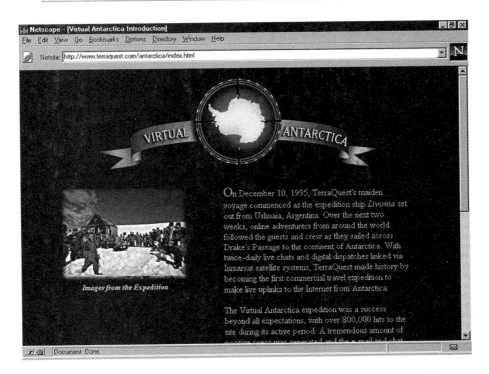

Many Internet sites describe different environments. The Virtual Antarctica Web page details the animals and plants of Antarctica.

Web site if you think you might want to become involved with the Jason Project with your teacher and class. The upcoming Jason X expedition will travel to the Peruvian Amazon rain forest.

LEO-15 Ocean Research Station

http://marine.rutgers.edu/pt/activities/cred.htm

LEO-15 is an underwater long-term ecological laboratory. This Web site provides middle-school-level Internet activities in ecology. You can visit this site to study coastal upwelling on the New Jersey shore. You can also study the Gulf stream and phytoplankton (microscopic sea plants).

One Sky, Many Voices

http://onesky.engin.umich.edu/

> *This site has four-week and eight-week environmental science programs that use the Internet and other technological communications tools. Different projects are featured all the time. If the timing is right, you might join one for your project. Global science topics are featured here.*

United States Environmental Protection Agency

http://www.epa.gov/

> *No chapter on environmental science would be complete without mentioning the United States Environmental Protection Agency. This federal organization is devoted to safeguarding our natural resources. Its home page has links for students, including very specific activities you can try at home and school. Start here, or jump directly to the*

Explorer's Club

http://www.epa.gov/kids/

> *where you will find activities on water quality and experiments on water filtration. This site was created by the United States Environmental Protection Agency.*

Water In The City

http://www.fi.edu/city/water/

> *You do not have to live on a farm to care or learn about water. Water In The City explains urban water supplies. Learn about reservoirs, sewers, water quality, and more, and then try a few of the water activities posted there.*

Other Environmental Science Sites

Air Pollution

http://www-wilson.ucsd.edu/education/airpollution/air pollution.html

> *This physical chemistry site teaches about sources and effects of air pollution. You can use it as a basis for a study of pollution where you live. For instance, you can place clear tape in various locations, sticky side exposed, for three days and then count, through a microscope, the number of particles that adhere per square centimeter. How do the levels of air pollution compare between the areas around a streetlight and in your bedroom? (Hopefully it is less inside!)*

Demographic Data Viewer Home Page

http://plue.sedac.ciesin.org/plue/ddviewer/

> *This is a mapmaking Web site that you can use whether or not your Web browser has Java. At this site, you can generate maps showing United States census information such as age, gender, income, race, and housing information. You can generate different maps of where you live to help show the environmental impact that population growth has on your community. It creates great graphics.*

Satellite Photographs and Imaging

KidSat

http://kidsat.jpl.nasa.gov/

> *KidSat is planned and operated by students who want to explore Earth from space. Via remote-sensing instruments*

that gather data and do research, students can zoom in to specific geographic regions with the click of a mouse. For instance, American kids have studied Indonesian fires from space.

Advanced Hydrology Sites

These sites are more advanced information sources on water. Your teacher may have ideas on how you can use the government-funded data available here.

Global Hydrology Resource Center
http://ghrc.msfc.nasa.gov/
> *This is a data-rich site with measurements on wind speed, water vapor, and other hydrology-related quantities.*

National Snow and Ice Data Center
http://www-nsidc.colorado.edu/
> *NSIDC provides data on snow and ice in digital form. It has an education section with questions and answers on ice and snow, glaciers, and avalanche awareness. It also contains a "mapping and gridding primer," which teaches about mapping. It defines such terms as points, pixels, grids, and cells. You can learn about how satellites are used to create maps.*

Oak Ridge National Laboratory
http://www-eosdis.ornl.gov/
> *This site provides remote-sensing data on wetlands, grasslands, and water bodies.*

Chapter 6

Astronomy Projects Using the Internet

Caitlin wanted to do an astronomy project for her seventh-grade science class. She had a telescope on her back porch that she used to view the night sky. At certain times of the year, she could see planets such as Venus and Mars. Planets orbit the sun, which is called their revolution. They also spin about their axes, which is called their rotation. For her project, Caitlin decided to compare the rotation and revolution of different planets to determine whether these properties might have an effect on the formation of life. Earth is the only planet that we know with certainty has life on it. It is one of the terrestrial planets. Venus and Mars are also terrestrial. They are similar to Earth in size. They were formed in the same way as Earth, by the gathering of clumps of matter. And they contain similar elements. Caitlin hypothesized that some planets have conditions that would prevent the formation of advanced life, but that Earth conditions, including rotation and revolution rates, support life.

Caitlin started her online research at

NASA Home Page

http://www.nasa.gov/

The National Aeronautics and Space Administration is a large federal agency devoted to space exploration. Caitlin noticed a "Search the NASA Web" link at the bottom of the home page. She clicked there and entered keywords such as "rotation" and "revolution." Many links came up. She learned about the planets by visiting these sites and reading the information and viewing the images found there.

Caitlin began to learn about the different features of planets and the ways scientists measure them. She used some additional planetary study sites in her project, including the following two:

Welcome to the Planets

http://pds.jpl.nasa.gov/planets/

The Nine Planets

http://www.seds.org/nineplanets/nineplanets/

These two sites provide good introductory information on the planets. Caitlin was able to click on links for each planet to get pictures and data. She used Microsoft Word® to make a table summarizing data she found on the Web. The purpose was to compare the planets. Here are the first few rows of her table:

PLANET	ORBIT RADIUS (AU)	PERIOD OF ROTATION (Days)	PERIOD OF REVOLUTION (Years)
Mercury	0.387	59.0	0.241
Venus	0.723	243	0.615
Earth	1.00	1.00	1.00

Already Caitlin was starting to get a picture of how the planets differ. Mercury is only a third the distance from the sun as Earth. It would be much too hot for life to exist there as we know it. She realized Venus, too, would be a poor place for life because of its long day. This long day, combined with a carbon dioxide atmosphere, would make the surface of Venus extremely hot. She concluded that a planet much different from Earth in its rotation and revolution would be unlikely to produce life similar to ours.

Caitlin wanted some images of the planets from outer space and at their surfaces. She found them at these sites:

MSNBC's Time & Again: Planetary Exploration
http://www.msnbc.com/onair/msnbc/timeandagain/archive/Solar/default.asp

Views of the Solar System
http://bang.lanl.gov/solarsys/

She discovered that there were thousands of images of planets on the Internet, and more and more were made available each day. She did not want to use too many images in

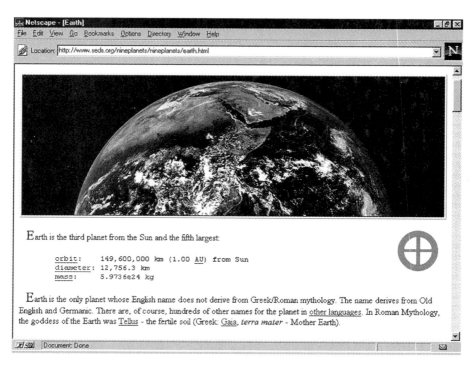

Facts about the planets, as well as images, can be found at the many different planet Web sites.

her project, so she selected three from outer space and three close-ups of the surfaces. She printed them in color to include in her report. She decided it would be best to go with a few large, detailed images rather than too many smaller ones.

Caitlin wanted to show the motion of the planets around the sun, so she used this online simulation:

Raman's Orbit Simulator—Gravitational Force
http://www.ppsa.com/science/orbit.htm

When she got to the Web site, it notified her that she needed Shockwave to run it. Her brother helped her follow

78

the directions for getting this plug-in. Then she was able to view moving images of planet paths and to control the simulation. This inspired her to observe the planets themselves from her home. She used her telescope to try to locate them. She got information from the Skywatch column in her local newspaper on where and when to look. Her own observations were an important part of her report.

Other Sites to Help You in Your Astronomy Research

Aurora Borealis

The Aurora Page
http://www.geo.mtu.edu/weather/aurora/

> *The aurora borealis, also known as the northern lights, is one of the most beautiful celestial events you can witness. Normally it is difficult to see the northern lights unless you live very far north, such as in Alaska. But on the Aurora Page, you can read the latest sighting reports for North America, the latest three-day forecast, and the latest three-hour reports.*

Auroras: Paintings in the Sky
http://www.exploratorium.edu/learning_studio/auroras/ index.html

> *Here is another site where you can find out how auroras are formed and why they have such brilliant colors.*

Comets and Other Astronomical Objects

Near Earth Asteroid Rendezvous

http://sd-www.jhuapl.edu/NEAR/

This is the home page for the NEAR project, which examines asteroids and other chunks of matter orbiting the sun in the neighborhood of Earth. Such studies shed light on Earth's history: how Earth and the other planets were formed; when Earth was struck by asteroids; and how life might have formed.

Small Planet Communications

http://www.smplanet.com/science/SL9.html

This site describes how you can investigate Comet Shoemaker-Levy 9, which smashed into the planet Jupiter in July 1994. It gives instructions on how to create your own Web page of research findings. For your project, you can gather astronomical data from the Internet and create an informative Web site. You will have your own place on the information highway.

General Astronomy Resources

Earth & Sky Radio Series Online

http://www.earthsky.com/

This site is devoted to the Earth & Sky Radio Series, which is all about astronomy, earth science, and environmental science. It has questions and answers, along with a searchable database. You can join the Earth & Sky mailing list at this site.

Observatorium

http://observe.ivv.nasa.gov/

> *NASA's Observatorium is a public access site for Earth and space data. It has pictures of the planets and stars, as well as the stories behind those images.*

Outer Orbit

http://www.outerorbit.com/

> *Space News' Outer Orbit Web site is an educational site addressing space explorations through classroom activities and links to other space-related sites on the World Wide Web. The featured lesson plans and archives are organized into grades 5–8 and 9–12; students can read them for project ideas. Also here: space chat, ask-an-astronomer, and space links.*

Solar System Visualization Project

http://www-ssv.jpl.nasa.gov/SSV/SSV_home.html

> *This project was created by NASA's Jet Propulsion Lab and other space agencies. You can study the red spot on Jupiter, asteroid flybys, and space missions. You will find images, movies, and links to other space sites here.*

Sources of Astronomy Information

http://heasarc.gsfc.nasa.gov/docs/outreach.html

> *This site is provided by the Laboratory for High Energy Astrophysics (LHEA). It is the gateway to Imagine the Universe, Starchild, Ask a NASA Scientist, and other student space Web sites.*

Windows to the Universe

http://www.windows.umich.edu/

> *This project is funded by NASA. You can enter the Web site in a variety of ways—using a CD-ROM, Frames, or no Frames.*

Mars

The Daily Martian Weather Report

http://nova.stanford.edu/projects/mgs/dmwr.html

> *At this site you can study the temperature and weather on Mars in detail.*

Images & Data from the Mars Pathfinder Mission

http://www.uflib.ufl.edu/

Mars Global Surveyor

http://www.jpl.nasa.gov/mgs

Mars Team Online

http://quest.arc.nasa.gov/mars/

Welcome to the Mars Pathfinder Mission!

http://mpfwww.jpl.nasa.gov/

> *These sites include data and images brought to us by unmanned spaceflights to our red neighbor. Mars is so much like Earth in its properties that scientists debate whether life exists there or might have in the past.*

The Moon and Lunar Landings

In the space race of the 1950s and 1960s, the United States and the Soviet Union competed to be the first country to put a man on the moon. The Web is filled with many resources that describe manned and unmanned missions to the moon and the features and properties of the moon itself.

Kennedy Space Center: Project Apollo

http://www.ksc.nasa.gov/history/apollo/apollo.html

Here you will find information on the famous Apollo moon landings. This site has information about the crews, the launches, and the journeys. It links to other Apollo sites at Kennedy Space Center and beyond. Other sites that relate to the Apollo programs include:

The Apollo Program

http://www.stockbridge.k12.wi.us/

The Apollo Program (1963–1972)

http://nssdc.gsfc.nasa.gov/planetary/lunar/apollo.html

Soviet Missions to the Moon

http://nssdc.gsfc.nasa.gov/planetary/lunar/lunarussr.html

These sites offer historic background, technical details of the flights, and scientific findings.

Kids on the Moon!

http://www.njin.net/~dmollica/index.html

Students of the Immaculate Conception School in New Jersey have placed their hydroponics project online to share with all. Their space garden design may be useful when people start to inhabit the moon and other planets.

Lunar Exploration

http://nssdc.gsfc.nasa.gov/planetary/lunar/apollo_25th.html

> *This is a good starting point for moon research. It gives the historical background behind moon exploration and has links to all the major moon expeditions.*

Moonlink

http://www.moonlink.com/

> *In January of 1998,* Lunar Prospector *was launched to gather data about the moon's surface. Moonlink is an Internet-based education program of NASA's* Lunar Prospector *mission. You can select a surface patch on the moon and zoom in for a closer look, and you can find out how other students are using the data from this mission.*

Surveyor to the Moon (1966–1968)

http://nssdc.gsfc.nasa.gov/planetary/lunar/surveyor.html

> *Before sending people to the moon, we sent unmanned crafts to take pictures and determine whether a visit would be safe for astronauts. This site has experiment descriptions for* Surveyor 1 *through* Surveyor 7.

Other Planets

3-D Tour of the Solar System

http://cass.jsc.nasa.gov/research/stereo_atlas/SS3D.HTM

> *Put on your 3-D glasses and see surfaces of planets in three dimensions. Special photography techniques make stereo space images available. Ordering instructions are provided here.*

The Galileo *Project*

http://nssdc.gsfc.nasa.gov/planetary/galileo.html

The Galileo *mission used the gravitational fields of Venus and Earth to send a spacecraft to Jupiter. When* Galileo *got there, it dropped a probe into the atmosphere of the huge gas planet. The probe sent information back to Earth. This Web site presents the most recent information from this mission.*

Liftoff to Space Exploration

http://liftoff.msfc.nasa.gov/

Figure out your age on Mars—and on other planets—with a special online calculator. Guess what? You are probably only about one year old on Jupiter. You can contribute a picture to the Kid's Space Art Gallery here.

A Solar System Scale Model Meta Page

http://www.tiac.net/users/mcharity/export0/solarsystem/

This site deals exclusively with the size and proportion of the solar system. Visit it if you need to create models of planet size and distance to support your project.

Space Travel

Maximov Online: The Mir *Space Station*

http://www.maximov.com/Mir/mir2.html

At Maximov Online you can learn about Russia's space ventures. You can link to information on the history and design of the Mir *Space Station.*

NASA Jet Propulsion Laboratory

http://www.jpl.nasa.gov/

NASA's rockets send our probes and people into space. The JPL site contains up-to-date information on spaceflights. It gives special attention to the recent Mars missions.

NASA Shuttle Web

http://shuttle.nasa.gov/index.html/

At this site you can chart the progress of current space missions. Find out about the crews and the planned activities. This site offers sneak previews of coming trips.

Stars and Nebulae

The Constellations

http://www.rose.com/~richard/

This Web site teaches about the constellations. Constellations are groups of stars that we see from Earth's surface. The stars are related mostly by the way they appear; our ancestors looked up in the night sky and saw shapes of animals and people. This site presents a combination of historical information, maps, and scientific information.

MWO Online Star Map

http://www.mtwilson.edu/Services/StarMap/

This interactive site from the Mount Wilson Observatory allows you to enter information such as date and location to get a map of the night sky. You can choose to show such features as constellations and meteor showers on your map.

STARCHILD

http://heasarc.gsfc.nasa.gov/docs/StarChild/StarChild.html

Stars have properties just like planets. They have size, mass, color, and other measurable features. Starchild is a learning center for young astronomers. It is a good source of information on stars for students just entering middle school, and it has a glossary of space-related terms.

Star Colors and Temperatures

http://zebu.uoregon.edu/~soper/Stars/color.html

This Web page explains why stars have different colors.

Star Journey

http://www.nationalgeographic.com/features/97/stars/

At this site, hosted by National Geographic, you can view a star map of the night sky. You can move around the night sky and zoom in on particular regions. You will also be able to see images from the Hubble Space Telescope by clicking on them within the map.

The Web Nebulae

http://seds.lpl.arizona.edu/billa/twn/

Nebulae are clouds of dust and gas that, when viewed through telescopes, are often colorful and beautiful. This site has a collection of images of nebulae.

The Sun

The 150-Foot Solar Tower at Mount Wilson Observatory

http://www.astro.ucla.edu/~obs/intro.html

The telescope at Mount Wilson Observatory is aimed at the sun. You would not want to view the sun directly, because the sun can damage your eyes. However, scientists observe the sun indirectly, using special optical and photographic techniques. Here you can find out about sun spots and magnetic activity.

Stanford Solar Center

http://solar-center.stanford.edu/

At the Stanford Solar Center, you can learn about eclipses and view eclipse images. This Web site has fun activities such as a scavenger hunt and poster contest.

Virtual Sun

http://www.astro.uva.nl/michielb/sun/kaft.htm

This Web site offers a twenty-minute virtual tour of the sun, complete with movies.

Telescopes and Astronomical Data

Space Telescope Science Institute HST Public Pictures

http://oposite.stsci.edu/pubinfo/Pictures.html

What is the big deal about the Hubble Space Telescope (HST)? It is simple: Because it is located in outer space, above Earth's atmosphere, there is no atmosphere or clouds to distort the images that it collects. This provides crystal-clear images not possible from Earth. This site contains Hubble images and tells how they were gathered.

Advanced Astronomy Sites

The Hopkins Ultraviolet Telescope
http://praxis.pha.jhu.edu/hut.html

UCSB Remote Access Astronomy Project Remotely Operated Telescope
http://www.deepspace.ucsb.edu/rot.htm

> *These are advanced telescope sites. Astronomers use different types of telescopes in their studies. The first telescopes were optical telescopes. Now we also rely on radio, infrared, and ultraviolet telescopes.*

Additional telescope Web sites are listed at these two directories:

Radio Telescope Resources
http://www.stsci.edu/astroweb/cat-radio.html

Yahoo: Science: Astronomy:Telescopes
http://www.yahoo.com/Science/Astronomy/Telescopes/

> *These sites lead to telescope Web sites around the world.*

Chapter 7

Chemistry Projects Using the Internet

Sylvia had become very interested in acids and bases. She had learned how to use litmus and pH paper to test the acidity of chemicals. She knew from science class that many substances at home are acids and bases. Vinegar and lemon juice are common household acids. Baking soda is a base. Water is neutral. Sylvia decided to do a more formal study of household acids and bases. She wanted to know, which substances are acids, which are bases, and which are neutral? Are there any patterns to be found in the chemicals at home?

To prepare for her project, she did a search for "acids and bases" at this student site:

Chem4Kids!
http://www.chem4kids.com/

This Web page came up:

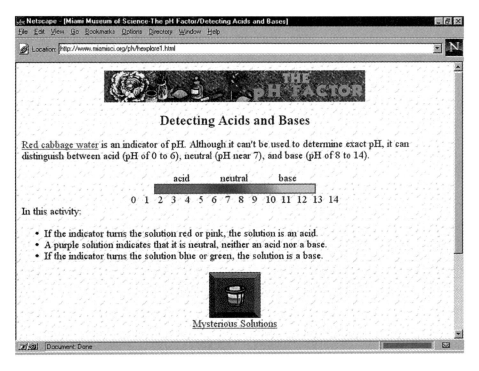

Chem4Kids Acids & Bases

http://chem4kids.com/reactions/acidbase.html

It described the pH scale for measuring acidity. This scale ranges from 0 (acidic) to 14 (basic), with 7 being neutral. It gave examples of acids and bases and their actual pH values. It defined important acid-base terms and explained the chemistry of acid-base reactions—all in simple terms. Another useful site she used was

The pH Factor

http://www.miamisci.org/ph/default.html

This site had interesting information on tasting nontoxic acids and bases. Acids taste sour, and bases taste bitter. Following the

Netscape - [Miami Museum of Science-The pH Factor/Detecting Acids and Bases]

File Edit View Go Bookmarks Options Directory Window Help

Location: http://www.miamisci.org/ph/hexplore1.html

THE pH FACTOR

Detecting Acids and Bases

Red cabbage water is an indicator of pH. Although it can't be used to determine exact pH, it can distinguish between acid (pH of 0 to 6), neutral (pH near 7), and base (pH of 8 to 14).

acid   neutral   base

0 1 2 3 4 5 6 7 8 9 10 11 12 13 14

In this activity:

- If the indicator turns the solution red or pink, the solution is an acid.
- A purple solution indicates that it is neutral, neither an acid nor a base.
- If the indicator turns the solution blue or green, the solution is a base.

Mysterious Solutions

Document: Done

The pH scale for measuring acids and bases can be used for your science project.

instructions there, Sylvia was able to verify this with her tongue, using small quantities of safe-to-taste chemicals such as citrus juices, baking soda, and antacid.

In the main part of her research, Sylvia measured the pH of different household substances, using pH paper she got from her teacher. She made a table of her data. Here are some of her findings:

SUBSTANCE	pH	CLASSIFICATION
lemonade	4	acid
apple juice	5	acid
egg white	8	base
milk	8	base
toilet bowl cleaner	1	strong acid
window cleaner	10	base
laundry detergent	9	base
ammonia	13	strong base
water	7	neutral
furniture polish	7	neutral
corn oil	7	neutral
cola	5	acid
stomach antacid	8	base

From the results of her experiment, Sylvia formed some important conclusions. Citrus juices and soft drinks are acidic. Oils and water are neutral. Cleansers and antacids are basic. Sometimes there were exceptions to these trends. Toilet bowl cleaner turned out to be acidic. She found hydrochloric acid in the list of ingredients, which explained this. In fact, many of the substances she studied had ingredients written on the packages. Citric acid, ammonium hydroxide, and acetic acid are examples of chemicals that she found influence the pH of many home products.

Sylvia took her research one step further to find out what makes something an acid or a base. Some useful definitions she got from the Web sites above were

Acids: Chemicals that donate a hydrogen ion, or H^+, when they react.

Bases: Chemicals that donate a hydroxide ion, or OH^-, when they react.

Hydrogen and oxygen are very important elements in chemistry. To learn more about these elements in particular, she visited a periodic table on the Web:

WebElements
http://www.shef.ac.uk/~chem/Web-elements/

WebElements had a virtual periodic table of the elements. Sylvia clicked her mouse on hydrogen to find out about this element. She learned its atomic weight, atomic number, and physical properties. She also learned its chemical importance and some historic information about it.

Other Sites to Help You in Your Chemistry Research

Chemicals

In the Lab of Shakhashiri

http://scifun.chem.wisc.edu/scifun.html

> Chemistry professor Bassam Z. Shakhashiri, at the University of Wisconsin–Madison, is the master of chemical demonstrations. He reveals his secrets at this site. A special section contains chemical experiments you can do at home.

K–12 Water and Ice Module

http://cwis.nyu.edu/pages/mathmol/modules/water/water_teacher.html

> This site contains information about ice. You will find laboratory simulations, molecular simulations, and concepts and challenges.

What You Always Wanted to Know About Salt

http://www.saltinstitute.org/4.html

> This site has information on one of the most famous chemicals of all: sodium chloride, otherwise known as table salt. It teaches about the chemical properties of salt, the history of salt, and uses for this important chemical.

Experiments

Chemistry Experiments You Can Do at Home

http://tqd.advanced.org/2690/exper/exper.htm

This site has many experiments you can do at home. Although it is directed at high school chemistry students, many of the activities use common materials and would not be difficult to do in your kitchen. Experiments include testing for vitamin C and growing salt crystals.

Periodic Tables

Chemicool Periodic Table
http://wild-turkey.mit.edu/Chemicool/

The MPEG Periodic Table
http://www.cm.utexas.edu/mpegtable.html

Periodic Table of the Elements
http://mwanal.lanl.gov/CST/imagemap/periodic/periodic.html

The Periodic Table of the Elements on the Internet
http://domains.twave.net/domain/yinon/default.html

These virtual periodic tables contain more information than printed tables. Go to these sites and you will see the natural and synthetic elements. Click on an element and detailed information will appear. You can get the element's atomic numbers, mass number, properties, chemical importance, and history. MPEG is a Web-based periodic table that includes videos of element reactions.

Molecules and Atoms

Gallery of Molecular Machines
http://www.wag.caltech.edu/gallery/gallery_nanotec.html

Netscape - [A Periodic Table of the Elements at Los Alamos National Laboratory]

File   Edit   View   Go   Bookmarks   Options   Directory   Window   Help

Location: http://mwanal.lanl.gov/CST/imagemap/periodic/periodic.html

Periodic Table of the Elements

Click an element above for more information.

How to use the Periodic Table

List of Elements and Their Symbols

Document: Done

A periodic table of the elements, many of which can be found on the Internet, is helpful for many chemistry projects.

GRASP Sample Images

http://tincan.bioc.columbia.edu/Lab/grasp/pictures.html

Molecular Art Gallery

http://www.wag.caltech.edu/gallery/art_gallery.html

These sites have beautiful color images of molecules. Some of the sites are advanced, but you may find the explanations and pictures useful if you are studying molecules.

Life, the Universe, and the Electron

http://www.nmsi.ac.uk/on-line/electron/

This Web site celebrates the one hundredth anniversary of the discovery of the electron. It tells us what electrons are, how we know about them, and what they mean to modern scientists.

MicroWorlds

http://www.lbl.gov/MicroWorlds/

Scientists use a machine called the advanced light source (ALS) to learn about atoms and molecules. In MicroWorlds you can learn how the ALS works and what it tells us about matter. You can register at the site and receive a free poster.

Chapter 8

Physics Projects Using the Internet

Eddie had a pendulum cuckoo clock in his room. He could adjust the speed of the clock by moving the weight on the pendulum. This made him curious. He wondered about the relationship between length and time for the pendulum. And what about other variables? How would changing the mass of the pendulum affect the speed of the clock? Since he had to complete a science fair project, he decided to construct an experiment to answer these questions.

To prepare for his research, Eddie searched the Internet for information that would relate to the pendulum. He visited the

How Things Work Search Page
http://howthingswork.virginia.edu/

to see whether anyone had asked about the pendulum before him.

He then did a search for "pendulum," and several questions and answers came up. Some were quite useful to him. One question and answer defined some important terms for him, such as period, amplitude, and frequency.

Eddie also found this pendulum-related site:

Simple Harmonic Motion
http://www.explorescience.com/harmonic.htm

At first he could not see graphics at this site. Then he followed the directions it gave for installing a plug-in on his computer called Macromedia Shockwave. What he then saw

The Visualize Science home page can lead you to interactive science project sites.

inspired him. There was a picture of a spring and a pendulum—one that moved. In this simulation, Eddie could change the stiffness and the mass of the spring, the length and the mass of the pendulum, and the force of gravity (g). After playing with the controls for a while, he started to realize that certain patterns existed. For instance, long pendulums move slower than short ones, but changing the mass of the pendulum does not affect its period (the time to swing back and forth once).

Eddie built an actual pendulum to explore the relationship between the length and period of the pendulum. He set the length of the pendulum to 0.2 meters. He timed how long it took for the pendulum to swing back and forth ten times, to the nearest tenth of a second. He set up a table for his data.

LENGTH (m)	10 SWINGS (s)	1 SWING (s)
0.20	17.5	1.75
0.40		
0.60		

DO NOT WRITE IN THIS BOOK.

The first column gives the length of the pendulum in meters. Eddie hung the pendulum from a pipe near the ceiling for the longer lengths. The second column is the time required for ten swings, which he recorded with a stopwatch to the nearest tenth of a second. The third column is simply the number in the second column divided by ten. Eddie's purpose of timing ten swings and then dividing by ten was to reduce

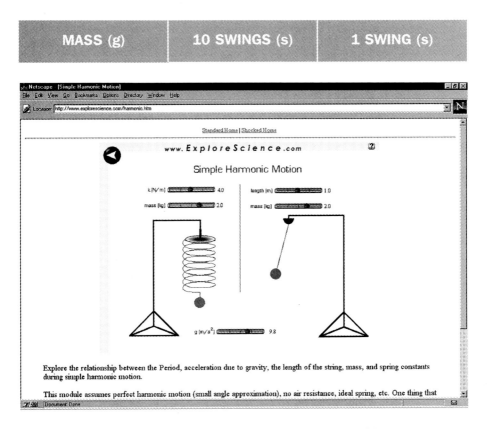
the human error in timing. He took ten measurements, from 0.2 m to 2.0 m, and completed the table. Then he graphed time as a function of length. His graph showed that as the pendulum got longer, the time it took to swing back and forth was also longer. When he made this type of adjustment on his cuckoo clock, it would run slower. This made sense after his experiment.

Eddie changed his experimental method. This time, instead of changing the length of the pendulum, he changed the mass and kept the length constant. His table headings looked like this:

MASS (g)	10 SWINGS (s)	1 SWING (s)

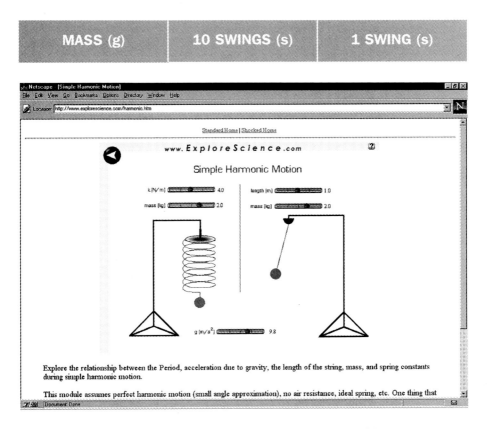

At ExploreScience.com, you can experiment with a pendulum.

What he found was a little surprising, but it agreed with the simple harmonic motion simulation: The period of the pendulum did not change with mass.

Eddie tried changing the amplitude of the swing—how far he pulled the pendulum to the side—to see how that affected the period. Amplitude, too, had no effect on the period.

Using his experimental data, combined with theories and historical information he found on the Web, Eddie determined that the only factors that affected the period of the pendulum are length and the gravitational pull of the earth. (Air resistance was a minor force and barely had any effect on the pendulum.) He also found that the pendulum and the spring move according to simple harmonic motion, which is common in nature and technology and important to scientists and engineers.

Other Sites to Help You in Your Physics Research

Density, Buoyancy, Fluids, and Air Forces

Air Travelers

http://www.omsi.edu/sln/air/

Air Travelers is a site dedicated to the science of hot-air balloons. You can learn about gas properties and buoyancy from the information and activities provided there. The site is designed for teachers, but you can probably sneak into the Balloon Activities section on your own to get some project ideas.

Dragonfly Flight Activities

http://www.muohio.edu/dragonfly/flight/flight_contents.htmlx

You can build paper airplanes and design a human-powered plane at this site.

NASA's Aerodynamics In Car Racing

http://www.nas.nasa.gov/NAS/Education/TeacherWork/RaceCar/Aerodynamics_In_Car_Racing.html

This site discusses the aerodynamics that engineers must understand when they design race cars. The Bernoulli effect results when air rushes over a surface. This important effect helps race car drivers control their vehicles. It also provides lift to airplane wings.

The Science of Ballooning

http://www.pbs.org/wgbh/nova/balloon/science/

This is another site dedicated to ballooning. It gives the history of ballooning and discusses atmospheric science. An excellent section describes the jet stream.

WaterWorks

http://www.omsi.edu/sln/ww/

WaterWorks is devoted to the science of fountains. Fountains are an artistic and practical part of our environment. WaterWorks gives instructions on investigating fountains and creating your own designs.

General Physics

Energy Quest

http://www.energy.ca.gov/education/index.html

Energy Quest is an award-winning site from the California Energy Commission. You can learn all about energy: the different forms of energy, uses of energy, and conserving energy. It also covers topics such as alternative-fuel vehicles and geothermal energy.

Explore Science Home Page

http://www.ppsa.com/science/

If you have Shockwave loaded on your computer, this is a fascinating site. There are simulations in almost every branch of physics, including mechanics, electromagnetism, optics, astrophysics, and waves.

Glenbrook South Physics Home Page

http://www.glenbrook.k12.il.us/gbssci/phys/phys.html

This site contains lessons, theories, and simulations. Explore the pages at this site to see if you can find a project that suits your interests. You can visit their multimedia physics studios to see animations of trains, falling objects, rockets, and satellites. There is a special physics projects area that includes studies of roller coasters, egg drops, traffic, potential energy, heat, acoustics, astronomy, sports, sailing, relativity, and flight.

How Things Work

http://Landau1.phys.Virginia.EDU/Education/Teaching/
HowThingsWork/

This is an ask-a-scientist area devoted to physics. It is hosted by Lou Bloomfield, author of How Things Work: The Physics of Everyday Life.

Nano World

http://www.uq.oz.au/nanoworld/scalemag.html

> *This site is all about scale and magnification. You will see the relative sizes of some familiar objects. Measurement is very important in physics, and length is one of the fundamental quantities.*

U.S. Navy Time Service Department

http://tycho.usno.navy.mil/frontpage.html

> *Time is another fundamental quantity in physics. At this site you can get the exact time from the United States Naval Observatory Master Clock. Find out how the clock works and what other functions are served by the Time Service Department.*

Mechanics and Engineering

Dragonfly Tools

http://www.muohio.edu/dragonfly/tools/index.htmlx

> *This site teaches about tools. It is not limited to physics, but you will find plenty of physics in the egg-drop area. Your egg-drop design should consider air resistance, gravity, and other forces of physics.*

Java Projectile Motion

http://medb.physics.utoronto.ca/Web/fun/JAVA/trajplot/ trajplot.html

> *This site requires a browser feature called Java. If you have Java installed on your computer, you can do simulations of projectile motion. You can vary the speed, angle, and wind*

to see where the projectiles land. You can print images of the different paths of projectiles and their graphs.

Racing & Roller Coaster Design Agency
http://dimacs.rutgers.edu/~dmollica/starttoys.html

This is an online design academy you can join with your class. You can work with others across the Internet to design and test roller coasters.

The Science of Cycling
http://www.exploratorium.edu/cycling/

This Exploratorium site deals with all aspects of the physics of cycling. You can learn about frame materials, friction, gears, aerodynamics, and design and racing strategies.

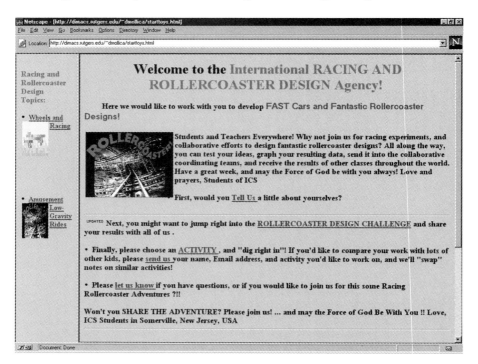

Design and test roller coasters using the principles of physics.

The Science of Hockey

http://www.exploratorium.edu/hockey/

The San Jose Sharks helped the Exploratorium create this exhibit. It deals with friction, reaction time, and collisions in this physics-filled sport. There is a Science of Hockey chat area, too.

Waves, Light, and Electromagnetism

About Rainbows

http://www.unidata.ucar.edu/staff/blynds/rnbw.html

What is a rainbow? How are they formed? What are the parts of a rainbow? When can you see them? The answers are at this award-winning site.

The Atoms Family

http://www.miamisci.org/af/sln/

The Atoms Family is a clever site about energy, particle physics, waves, electricity, and other topics in theoretical physics. Visit Dracula's Library, Frankenstein's Lightning Laboratory, and other scary rooms.

Bob Miller's Light Walk

http://www.exploratorium.edu/light_walk/index.html

Bob Miller is an artist, and the San Francisco Exploratorium hosts an online exhibit for him here. Activities are also provided. You can do your own "light walk," make a pinhole camera, and try slide projector activities.

The Electric Club

http://www.schoolnet.ca/math_sci/phys/electric-club/index.html

The Electric Club has thirty-seven experiments you can do on electricity with adult supervision. Examples of experiments include Make an Electromagnet, Joystick Switches, and Detecting Meteors.

Internet Museum of Holography

http://www.enter.net/~holostudio/

Enter the museum for educational resources on holograms. You will find a chat room, hologram newsletter, cyber theater, holo-games, and holo-links. You can learn what holograms are and how they are made.

Making Waves: An Online Guide to Sound and Electromagnetic Radiation

http://www.li.net/~stmarya/stm/home.htm

This is an online science project completed by St. Mary's School in Manhasset, New York. You can explore all the different wavelengths of radiation and sound. Learn about light, microwaves, X rays, and other forms of radiation and wave energy.

The Particle Adventure

http://pdg.lbl.gov/cpep/adventure.html

This site leads you through a tour of subatomic particles and explains the historical development of theoretical physics. Use the resources here to do a project that teaches how scientists discovered what they know about the organization of matter.

You now have the tools to support a successful science project using the Internet. Consider yourself special because you are a member of the first generation who used the World Wide Web in research and communication. One day you will rely on electronic resources professionally, no matter what career you choose. You will look back to the time when you learned Internet skills in school. Good luck for a safe and successful science project.

Further Reading

Frazier, Deneen, Barbara Kurshan, and Sara Armstrong. *Internet for Your Kids.* San Francisco, Calif.: Sybex, 1997.

Kazunas, Charnan, and Thomas Kazunas. *The Internet for Kids.* Danbury, Conn.: Children's Press, 1997.

Lampton, Christopher. *Home Page: An Introduction to Web Page Design.* Danbury, Conn.: Franklin Watts, Inc. 1997.

Levine, John R. and Carol Baroudi. *The Internet for Dummies.* Second Edition. Foster City, Calif.: IDG Books Worldwide, Inc., 1994.

Mitchell, Kim. *Kids on the Internet: A Beginners Guide.* Grand Rapids, Mich.: Instructional Fair, 1998.

Moran, Barbara, and Kathy Ivens. *Internet Directory for Kids & Parents.* Foster City, Calif.: IDG Books Worldwide, 1998.

Pedersen, Ted, and Francis Moss. *Internet for Kids! A Beginner's Guide to Surfing the Net.* New York: Price Stern Sloan, Inc., 1997.

Polly, Jean Armour. *Internet Kids & Family Yellow Pages.* Third edition. Berkeley, Calif.: Osborne McGraw-Hill, 1998.

Snedden, Robert. *The Internet.* Chatham, N.J.: Raintree Steck-Vaughn Publishers, 1998.

Index